Topics in Mining, Metallurgy and Materials Engineering

Series Editor

Carlos P. Bergmann, Federal University of Rio Grande do Sul, Porto Alegre, Rio Grande do Sul, Brazil

"Topics in Mining, Metallurgy and Materials Engineering" welcomes manuscripts in these three main focus areas: Extractive Metallurgy/Mineral Technology; Manufacturing Processes, and Materials Science and Technology. Manuscripts should present scientific solutions for technological problems. The three focus areas have a vertically lined multidisciplinarity, starting from mineral assets, their extraction and processing, their transformation into materials useful for the society, and their interaction with the environment.

More information about this series at http://www.springer.com/series/11054

Seiji Katayama

Fundamentals and Details of Laser Welding

Springer

Seiji Katayama
Joining and Welding Research Institute
(JWRI)
Osaka University
Osaka, Japan

ISSN 2364-3293 ISSN 2364-3307 (electronic)
Topics in Mining, Metallurgy and Materials Engineering
ISBN 978-981-15-7932-5 ISBN 978-981-15-7933-2 (eBook)
https://doi.org/10.1007/978-981-15-7933-2

This Springer imprint is published by the registered company Springer Nature Singapore Pte Ltd.
The registered company address is: 152 Beach Road, #21-01/04 Gateway East, Singapore 189721, Singapore

Preface

The word "laser" named by Mr. Gordon Gould, USA, is the acronym of "Light Amplification by Stimulated Emission of Radiation". Laser is artificially made light of electromagnetic waves. A laser beam can be focused to a small spot through lenses or mirrors, and consequently becomes an extremely high power-density heat-source. It can easily heat, melt, and/or evaporate any metals, ceramics, plastics, and so on. As a result, lasers are appropriately used in various industries as materials processing such as welding, brazing, soldering, drilling, cutting, marking, cleaning, cladding, additive manufacturing or 3-D printing, and so on. Especially, laser welding is receiving the highest attention among the laser materials processing.

Researches and developments of laser welding are performed all over the world. Up to about 1990, the developments in high-power CO_2 lasers and practical researches were actively carried out in advance in USA, and Japan, Germany, UK and France followed. In around 2000, 3 sets of 45 kW high power CO_2 laser and 2 sets of 10 kW YAG laser were introduced in the research institute and the industrial field of Japan, and a variety of researches and developments of laser materials processing were actively performed. In this century, high power disk, fiber and diode lasers have been developed. Germany has become the most advanced country in the laser materials processing and a leader of the world, and the following countries are China, USA, UK, Japan and France. On the other hand, the author has performed various welding processes such as laser welding and hybrid welding of aluminum alloys and HT steels, the effect of pulse-shaping on laser welding results, elucidation of microstructure of laser rapidly solidified zones, laser welding under zero-gravity, laser lap or butt welding and laser brazing of Zn-coated steel, laser welding or joining of dissimilar metals, metal-plastic and metal-CFRP, laser welding in vacuum, laser welding of thick steel plates, laser welding of titanium, stainless steels, mechanism elucidation of laser welding defects and development of preventive procedures, tracking, in-process monitoring, adaptive control and feedback repairing during laser welding, and so on, and has revealed various fundamental phenomena during welding with pulsed and CW lasers. The results obtained in the above-mentioned researches are described in this book.

Industrial lasers used for actual materials-processing have been developed up to now. Chapter 1 deals with such lasers used practically for joining and welding. High power lasers were developed first in CO_2 laser, followed by YAG laser. Thereafter, excimer lasers were developed, but these were mainly used for drilling, etc., but not for welding. Then high-power diode lasers were developed. Moreover, high power, high quality and high efficiency disk and fiber lasers were developed and were widely used for welding and cutting. In recent years, ultrashort pulse lasers such as picosecond and femtosecond lasers have been used for joining of glasses. And second harmonic lasers emitting green light from disk lasers and blue diode lasers have actively been developed for welding of copper sheets or plates.

In Chap. 2, fundamentals of laser-materials interaction are discussed since these should be understood in laser materials processing, especially welding. Chapter 3 describes features of laser welding and important welding results obtained with various practical lasers. Chapter 4 deals with welding phenomena during spot welding with pulsed YAG laser and bead welding with CW CO_2, YAG, disk or fiber lasers. In Chap. 5, formation mechanisms and preventive procedures of laser welding defects are discussed. Characteristic welding processes are explained in Chap. 6. In Chap. 7, process monitoring, in-process monitoring, seam-tracking and adaptive control during laser welding are written in detail. Chapter 8 describes different features of laser welding of various materials. Chapter 9 deals with laser welding or joining results of dissimilar materials. Chapter 10 describes features of industrial applications of laser welding or laser-arc hybrid welding.

A variety of laser welding results, and important knowledge and information are summarized in this book. Thus, this book should be good as a textbook of laser welding for university senior students and graduate students. The author really hopes that this book will contribute to all the readers of this book, researchers, students, experts, scientists, professors, makers and dealers handling laser apparatuses and the peripheral devices, and so on. This book will contribute to the further advance in laser welding.

The author would like to thank Mr. Shinichi Koizumi and Ms. Nobuko Hirota at Springer Japan, and Mr. Praveen Anand and Ms. Kokila Durairaj at Springer Nature for their encouragement, help and hard work.

Osaka, Japan

Dr. Seiji Katayama
Professor Emeritus, Osaka University
Director of Nadex Laser R&D Center

Contents

Chapter 1
Kinds and Characteristics of Lasers for Welding

1.1 Fundamentals of Laser

The word "laser" is the acronym of "Light Amplification by Stimulated Emission of Radiation" describing the physical process of light irradiation or emission principle. Laser is not naturally existent on earth but is artificially made light of electromagnetic waves.

The features of laser are summarized in Table 1.1 [1]. The laser possesses several excellent characteristics such as supreme monochromaticity or an extremely narrow frequency spectrum range, super-directivity or parallelism, strong spatial and temporal coherency, high focusing ability to a small spot, high intensity or high power density, instant concentration of light energy, and so on. A laser beam can be focused to a small spot through lenses or mirrors, and consequently becomes an extremely high power density (intensity) heat source. It can heat, melt, and/or evaporate metals, ceramics, plastics, etc., and can facilitate the processing of hard and/or brittle materials which are difficult to be processed with tools.

A laser beam can be delivered without attenuation in vacuum or in various atmospheres and can process any materials or even large-scale structures in a non-contact manner from a far-away location. The laser can process plastics or glasses and the inside of transparent matters. The processing results are not affected by magnetism of materials, which affects the processing with an electron beam. The lasers of short wavelengths possessing high photon energies can facilitate the dissociation or ionization of atoms or molecules and the chemical reaction of matters. Consequently, drilling or photochemical reaction can be achieved by such lasers. Moreover, lasers of short wavelengths and lasers of ultrashort pulse durations and ultrahigh peak power densities can perform ablation-processing, which signifies non-thermal evaporation or vaporization.

By utilizing the above features, lasers are used in various industrial fields as materials processing such as welding, brazing, soldering, drilling, cutting, marking, cleaning, cladding, additive manufacturing or 3D printing, martensite transformation

S. Katayama, *Fundamentals and Details of Laser Welding*,
Topics in Mining, Metallurgy and Materials Engineering,
https://doi.org/10.1007/978-981-15-7933-2_1

Table 1.1 Features of laser

Item	Natural light	Laser	Applications
① Mono-Chro-maticity	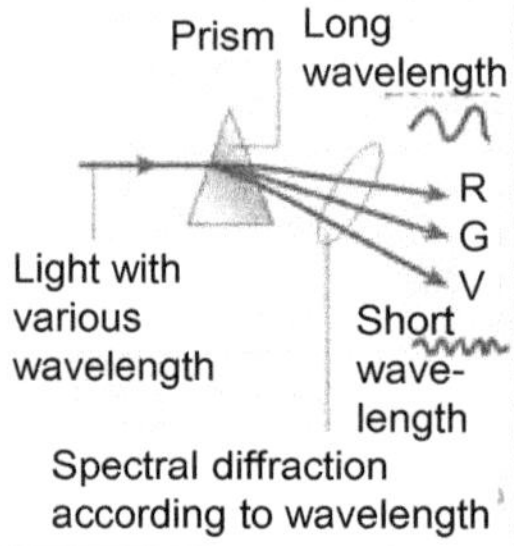 Prism; Long wavelength; R; G; V; Light with various wavelength; Short wave-length; Spectral diffraction according to wavelength	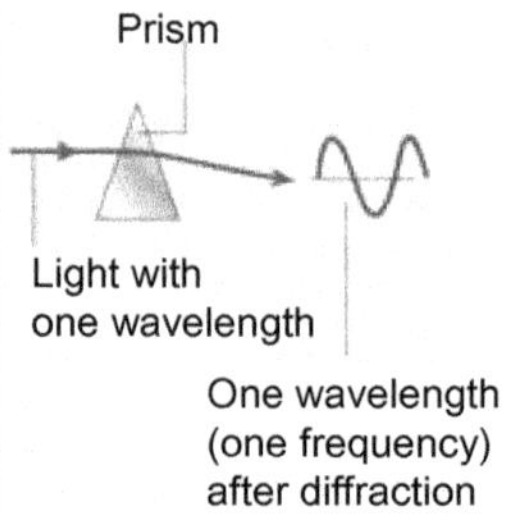 Prism; Light with one wavelength; One wavelength (one frequency) after diffraction	Spectroscopic analysis Isotope separation
② Direc-tivity	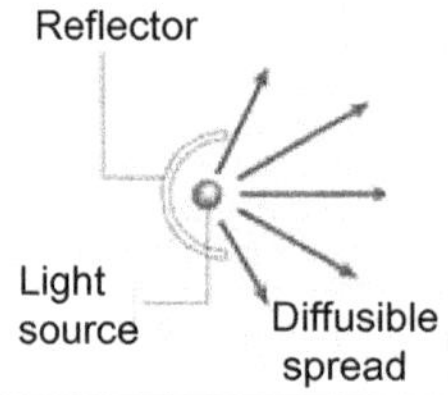 Reflector; Light source; Diffusible spread	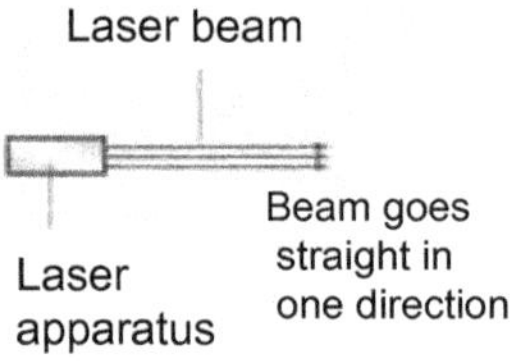 Laser beam; Laser apparatus; Beam goes straight in one direction	Optical communication Laser scanner Optical disk Laser radar
③ Cohe rency	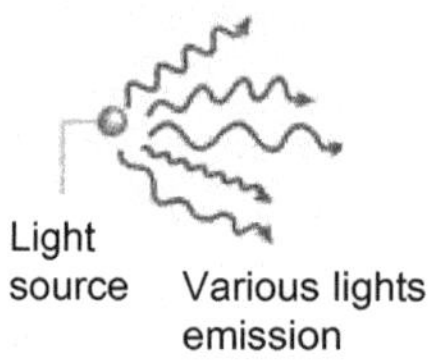 Light source; Various lights emission	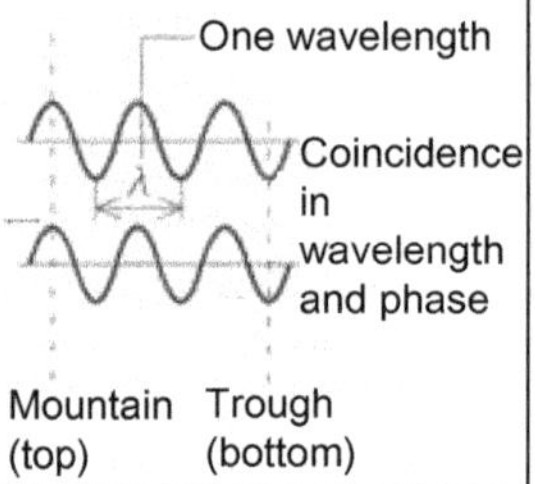 One wavelength; Coincidence in wavelength and phase; Mountain (top); Trough (bottom)	Holography Precision measurement due to interference fringe
④ Focusing/ High intensity (High power density)	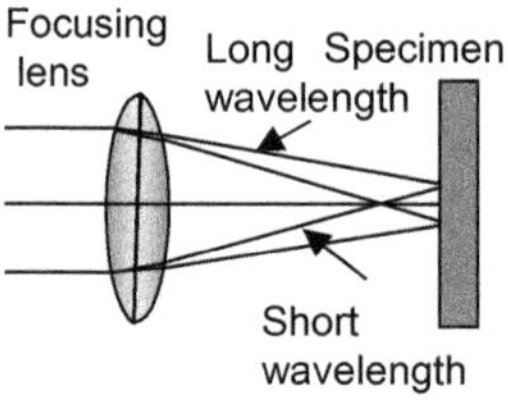 Focusing lens; Long wavelength; Specimen; Short wavelength	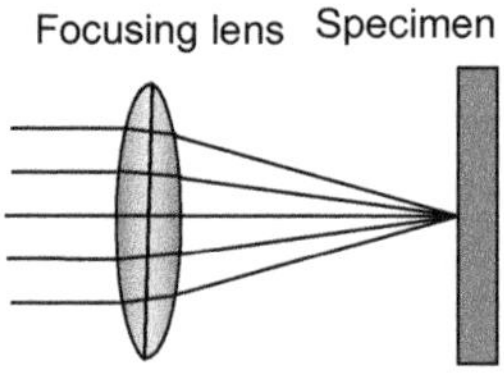 Focusing lens; Specimen	Laser processing Laser scalpel Laser weapons

hardening due to quenching, and so on [2]. Especially, laser welding is receiving the highest attention among the laser materials processing.

1.2 Principle of Laser Emission

The concept of energy levels is important for laser emission, which takes place via transition between different energy levels of an atomic or molecular system. The interaction of light (laser) with tiny matter such as atoms, ions, and molecules is understood by considering three different processes: absorption, fluorescence or spontaneous emission and stimulated emission, as the relationship between light and atomic states or energy levels is schematically shown in Fig. 1.1 [1, 3]. In the case of absorption, an atom in a low energy level is raised to a higher one by absorbing the light energy ($E = h\nu$; where E: light energy, h: Planck's constant, and ν: frequency). In fluorescence or spontaneous emission, an atom in an upper energy level decays spontaneously to a lower energy level and emits a photon with the appropriate frequency to satisfy the difference in energy. In stimulated emission, an atom in an upper energy level interacts with incoming light with the frequency corresponding to the energy difference, and the atom is stimulated to emit light by dropping to a lower energy level. Consequently, two photons emerge and have the same frequency, the same direction of travel and the same phase for the associated electric field. The light intensity is amplified, and thus, this process is the origin of laser emission.

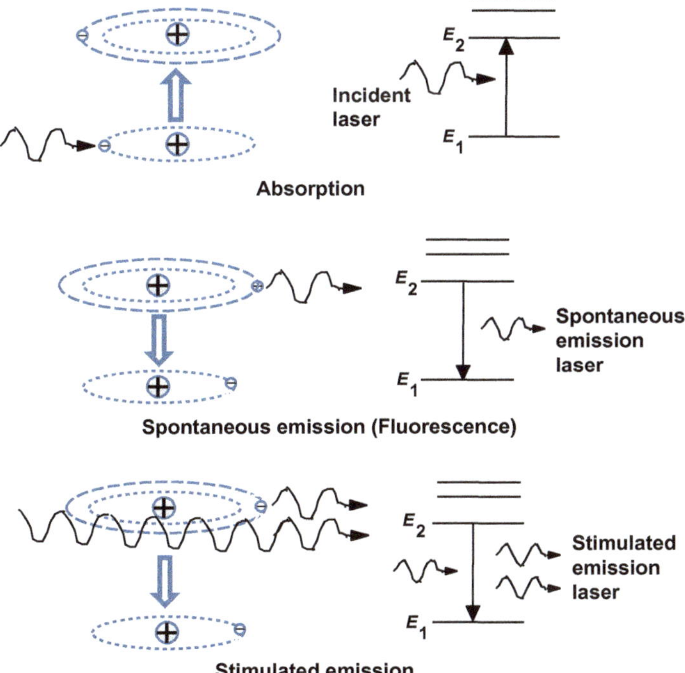

Fig. 1.1 Schematic illustration of absorption, spontaneous emission (fluorescence), and stimulated emission

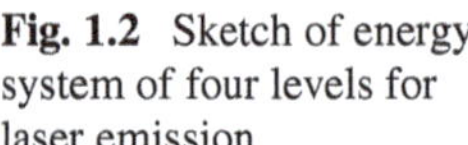

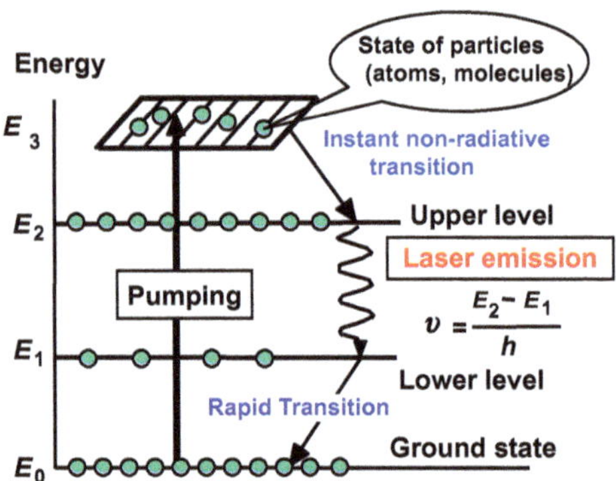

Fig. 1.2 Sketch of energy system of four levels for laser emission

Actually, the laser emission should require the utilization of energy systems with more than three levels and the use of an optical resonator or a resonant cavity with more than two reflection mirrors. The energy system of four levels for laser emission is sketched in Fig. 1.2 [1, 4]. The particles such as atoms or electrons in the base level are raised or pumped to the highest levels by the outer heat source such as light, plasma, electron beams, etc. These excited particles transit instantly and spontaneously to the upper level without emission. The particles in the lower energy level transit instantly to the base level in the similar way. Consequently, population inversion of particles takes place between the upper and the lower levels, and thereby the laser with the frequency (ν) corresponding to the energy difference ($E_2 - E_1$) is emitted.

Light can be reflected back and forth many times by the mirrors and is amplified by stimulated emission on each pass through the active medium. As an example of a resonant cavity with two mirrors, excitation and stimulated emission in a solid laser rod is schematically represented in Fig. 1.3 [4]. The original unexcited material is excited via optical pumping, followed by spontaneous emission or fluorescence and then stimulated emission. Emitted photons are repeatedly traveled between mirrors of total reflection (100% or 99%; for example, in the case of 99% reflection mirror, 1% is used for monitoring of laser power) and the lower value in reflection (e.g., 50%), and the photon intensity is increased by stimulated emission to properly radiate a laser beam through the one side mirror of the lower reflectivity.

In conclusion, it is important to know the necessity of more than three energy levels of active medium and the setting of more than two mirrors in the resonant cavity for laser emission.

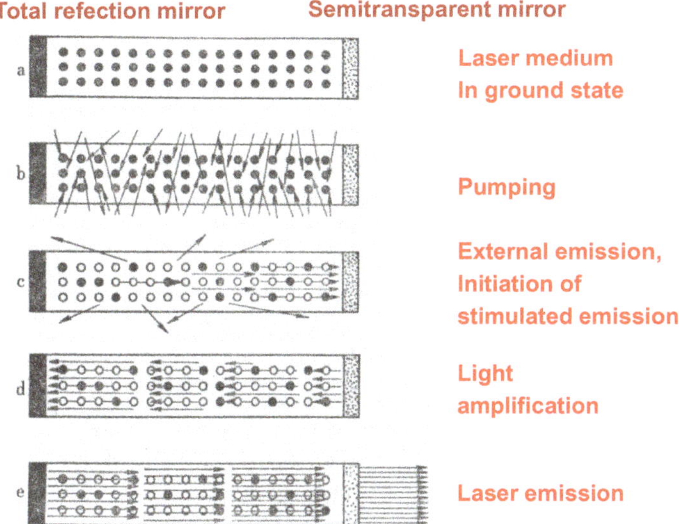

Fig. 1.3 Schematic representation of resonant cavity with two mirrors showing laser emission through excitation and stimulated emission in solid laser rod

1.3 Kinds and Characteristics of Lasers

Industrial lasers used for actual materials processing have been developed by many researchers and laser makers. Respective lasers are named according to the characteristics of emitted medium or material, shape, oscillation medium, wavelength, and pulse width, as shown in Table 1.2 [2, 5–8]. The lasers are understood by classifying into gas, liquid, and solid lasers. The solid lasers are the best in terms of easier operation. In particular, YAG laser of 1.06 μm wavelength, diode laser of about 0.5–2.2 μm wavelength, disk laser of 1.03 μm wavelength, and fiber laser of 1.07 μm wavelength can be delivered through optical fiber from several meters to about 200 m distance.

High-power lasers were developed first in CO_2 laser, followed by YAG laser. Thereafter, excimer lasers such as XeCl (308 nm), KrF (238 nm), and ArF lasers (193 nm) were developed, but these were mainly used for drilling or cutting of plastics, glass, etc., but not for welding. Then high-power diode lasers of about 0.8–1.1 μm in wavelength were developed and called direct diode lasers (DDLs) or high-power direct diode lasers (HPDDLs) as heat sources of materials processing. Moreover, high-quality high-power diode lasers were developed as pumping sources for disk and fiber lasers. In twenty-first century, high-power, high-quality, and high-efficiency disk and fiber lasers were developed and were widely used for welding and cutting. In recent years, ultrashort pulse lasers such as picosecond and femtosecond lasers, second harmonic lasers emitting green light from YAG, disk and fiber lasers, and blue diode lasers were actively developed. Ultrashort pulse lasers were chiefly

Table 1.2 Names and characteristics of lasers for welding or joining

> **CO_2 laser** (Wavelength: 10.6 μm; Far-infrared ray)
Laser media : CO_2–N_2–He mixed gas (Gas)
Average power [CW] : 45 kW (Maximum)
(Normal) 0.5–12 kW
Merit : Easier high power (Efficiency: 10–20%)

> **Lamp-pumped YAG laser** (Wavelength: 1.06 μm; Near-infrared ray)
Laser media : Nd^{3+}:$Y_3Al_5O_{12}$ garnet (Solid)
Average power [CW] : 10 kW (Cascade type Max & Fiber-coupling Max)
(Normal) 50 W–5 kW (Efficiency: 1–4 %)
Merit : Fiber-delivery and easier handling (good flexibility)

> **Diode laser (LD)** (Wavelength: 0.8–1.0 μm; Near-infrared ray)
(Blue laser) (Wavelength: 0.45 μm; Blue)
Laser media : InGaAsP, Al(In)GaAs, etc. (Solid)
Average power [CW] : 60 kW (Fiber-delivery Max.); 2 kW (Blue Max.)
Merit : Compact, fiber-delivery and high efficiency (20–50 %)

> **LD-pumped solid-state laser** (Wavelength: 1.06 μm; Near-infrared ray)
Laser media : Nd^{3+}:$Y_3Al_5O_{12}$ garnet (Solid), etc.
Average power [CW] : 13.5 kW (Fiber-coupling Max.)
[PW] : 6 kW (Slab type Max.)
Merit : Fiber-delivery, high brightness and high efficiency (10–20 %)

> **Disk laser** (Wavelength: 1.03 μm; Near-infrared ray)
Laser media : Yb^{3+}:YAG or YVO_4 (Solid), etc.
Average power [CW] : 16 kW (Cascade type Max.)
Merit : Fiber-delivery, 4 branches, high brightness high efficiency (15–25 %)

> **Fiber laser** (Wavelength: 1.07 μm; Near-infrared ray)
Laser media : Yb^{3+}:SiO_2 (Solid), etc.
Average power [CW] : 120 kW (Fiber-coupling Max.)
10 kW (single mode fiber laser Max.)
Merits : Compact, fiber-delivery, high brightness, high efficiency (20–50 %)

> **Green laser** (Wavelength: 515, 532, 535 nm; Green)
Laser media : Nonlinear optical crystal from disk, YAG or fiber laser
(Wavelength conversion element: LBO, KTP) (Solid)
Average power [CW] : 3 kW (Fiber-delivery Max.) for disk laser
[PW] : 500 W (Fiber-delivery Max.) QCW fiber laser
Merit : Short wavelength, high absorption for copper (5–20 %)

> **Ultrashort pulse laser** (Wavelength: 0.8–1.6 μm; Near-infrared ray)
Laser media : Titanium-sapphire, fiber laser and disk laser (Solid)
Average power [CW] : 1 kW (Max.) (Normal) 1–100 W
Merits : Ultrashort pulse, extremely narrow HAZ

used for drilling but are currently employed for lap-joining glasses. Green and blue lasers are expected to stably weld copper (Cu) plates or sheets, since the absorptance of these lasers to materials such as Cu and Au (gold) is high at room temperature.

1.4 CO_2 Laser (Carbon Dioxide Laser)

The in-cavity gases used for CO_2 laser emission are usually four mixed ones of CO_2, N_2, He, and CO. The gases are subjected to glow discharge under the pressure of 0.1 atm or less. The energy levels for the emission of CO_2 laser are shown in Fig. 1.4 [9]. CO_2 laser of 10.6 μm wavelength is emitted through the transition from the upper energy level of asymmetric stretching vibration to the lower energy level of symmetric stretching vibration of O–C–O triatomic molecules. N_2 gas helps to increase the number of CO_2 molecules at the upper levels. He gas can stabilize the laser power by stabilizing glow discharge. A proper amount of CO gas is effective to prevent power loss and unstable discharge. H_2O or moisture should be removed from the gases because it reduces laser power. Besides, CO_2 laser of 9.6 μm wavelength can be emitted by selecting reflectivity of mirrors in the transition to the lower energy level of bending vibration. This laser is chiefly used under the pulsed conditions for drilling of printed circuit boards [1].

A CO_2 laser material processing machine is schematically shown in Fig. 1.5 [1]. A laser beam from an resonate cavity travels in the space is bent by the mirror and is focused to a local small area by a lens or a mirror to produce a heat source of high power density or high intensity. The laser oscillator is called gas-sealed off type, low-speed axial flow type, high-speed axial flow type or 2-/3-axis orthogonal type

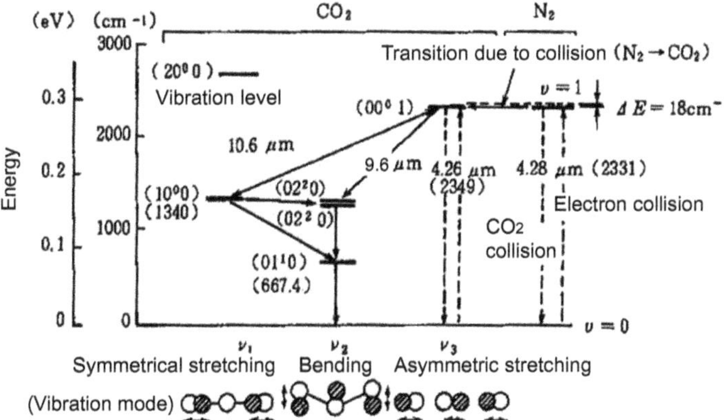

Fig. 1.4 Energy levels for CO_2 laser emission

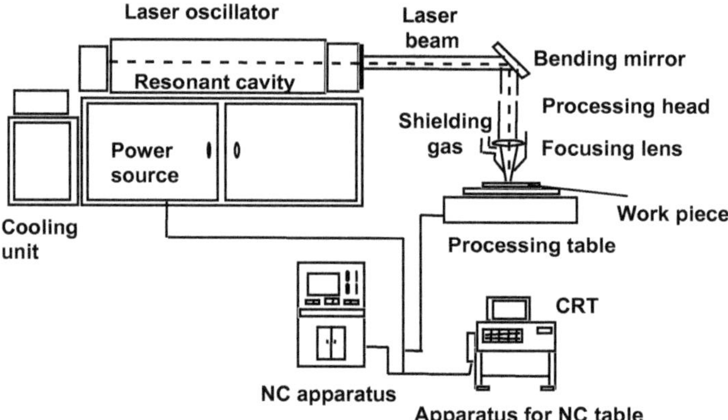

Fig. 1.5 Schematic arrangement of CO_2 laser machine

from the direction and characteristics of laser irradiation axis, gas flow and discharge current. CO_2 lasers are operated both in the continuous wave (CW) mode and in the pulsed wave (PW) mode. The lasers have different power density profiles of gauss distribution, low-order multimode, high-order multimode, donut mode, and so on. High-peak-power PW lasers are used for drilling, and high-power CW lasers are mainly employed for cutting and welding. In the case of CO_2 laser, ZnSe is used for transparent window and focusing lens. It should be noted in actual processing that the "thermal lens effect", under which a focal point of a focusing lens is normally shifted toward the lens due to non-uniform temperature distribution after laser passing, occurs easily.

Higher power was readily achieved in CO_2 lasers, and 10–45 kW class lasers were commercially available in the earliest stage. In welding with high-power lasers of more than 10 kW, He gas is used owing to no formation of a plasma, but on the other hand, in Ar shielding gas, a gas plasma is easily formed and the weld penetration becomes shallower. In welding of Al alloys with 2–4 kW CO_2 lasers in Ar shielding gas, an Ar plasma formed may sometimes facilitate melting and welding of Al alloys. Several kW high power lasers were practically used not only for welding of pulleys, motor cores, gears, etc. tailored blank welding of thin steel sheets, remote welding of parts in the automotive industry, but also for welding of the skins to stringers of aluminum alloys in aircraft industry and thick steel plates in the steel and shipbuilding industries. In recent years, such welding has been performed with a fiber, disk, or diode laser in place of CO_2 laser.

1.5 YAG Laser

YAG laser is named from the laser emitted from yttrium aluminum garnet (YAG; $Y_3Al_5O_{12}$) rod doping active ions of rare earth elements such as Nd^{3+} and Er^{3+}. It is normally Nd^{3+}: YAG laser of 1.06 μm wavelength. A YAG rod is the size of about 10 mm diameter and about 100–150 mm length. Light pumping is performed by arc lamps, flash lamps, or diode lasers. Nd^{3+}: YAG laser is emitted from the system of four energy levels. YVO_4 laser and YLF laser are in the YAG laser family.

A schematic arrangement of a YAG laser apparatus is shown in Fig. 1.6 [1]. Processing such as welding is performed by using direct optical system or fiber transmission optical system. Generally, flexible fiber delivery is almost always used in the materials processing because of a low transmission loss for a laser beam of 1.06 μm wavelength. GI (graded index)-type fibers are used for lasers with the powers of less than 500 W because of low damage threshold, although a higher-quality beam can be delivered. On the other hand, SI (step index)-type fibers are employed for higher-power lasers.

YAG lasers of different pulse shapes or continuous wave (CW) are emitted. Continuous wave (CW), pulsed wave from CW laser, pulse-modulated wave from CW YAG laser, normal pulsed wave, and Q-switched wave can be emitted under the conditions of different peak powers and pulse widths, as shown in Fig. 1.7 [1]. Normal pulsed YAG lasers, whose arbitrary pulse shapes can be set freely, have been used for welding of small parts such as battery cases, electric parts, glass flames, and ornaments. Besides, lamp-pumped 2–4 kW CW YAG lasers have been employed for tailored blank butt joints and lap joints of Zn-coated steel sheets. And, 6–10 kW CW YAG lasers have been utilized to produce deeply penetrated welds in stainless steel plates. Moreover, LD (laser diode) pumped YAG lasers have been used for three-dimensional welding of car roofs, stitch welding of car sills and partly remote lap welding of Zn-coated steel sheets. Nevertheless, at present, such lamp-pumped and LD-pumped high-power CW laser apparatuses are not manufactured because

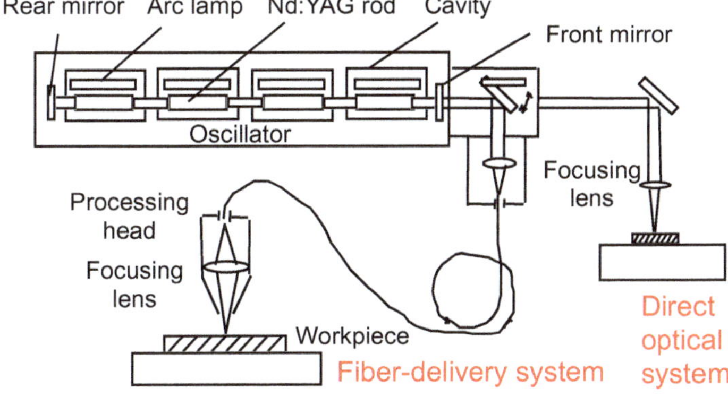

Fig. 1.6 Schematic arrangement of YAG laser machine with fiber delivery and direct optical systems

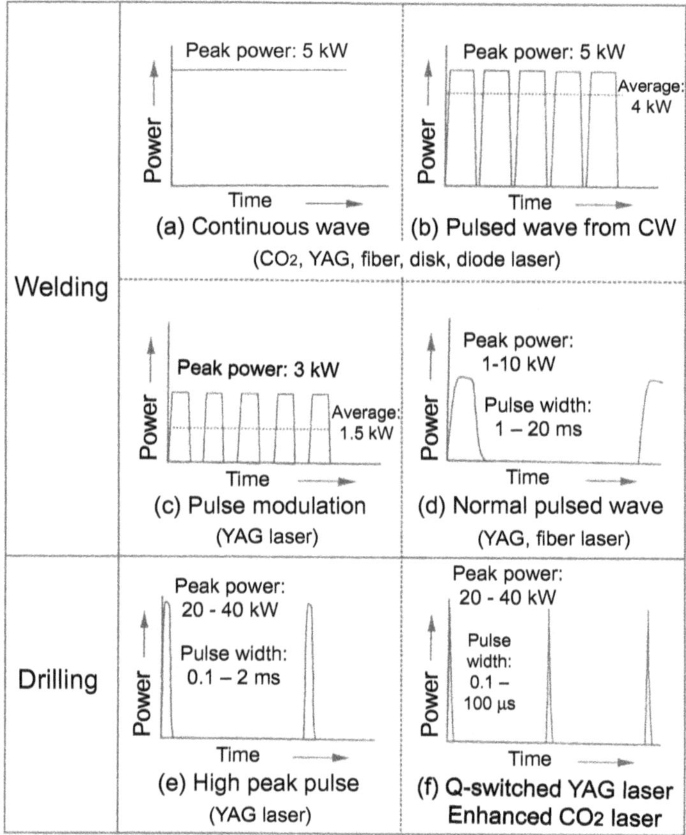

Fig. 1.7 Various characteristic wave shapes of CW laser and pulsed lasers for welding and cutting

of low lasing efficiency of 1–3% and less than 10%, respectively. Instead of these high-power CW YAG lasers, fiber, disk, or diode lasers are now in general use.

1.6 Diode Laser

In diode laser, a current is directly flown in the forward direction against a semiconductor of double hetero-junction structure from the outer side, and then the light is emitted by recombing electrons and positive holes at the active layer between an n-type semiconductor region and a p-type semiconductor region, as shown in Fig. 1.8 [10]. The photons move forth and back by the mirror and/or refraction rating, and the laser radiates due to the amplification by stimulated emission. Diode laser is called semiconductor laser or laser diode (LD in short). In this book, the name of "diode

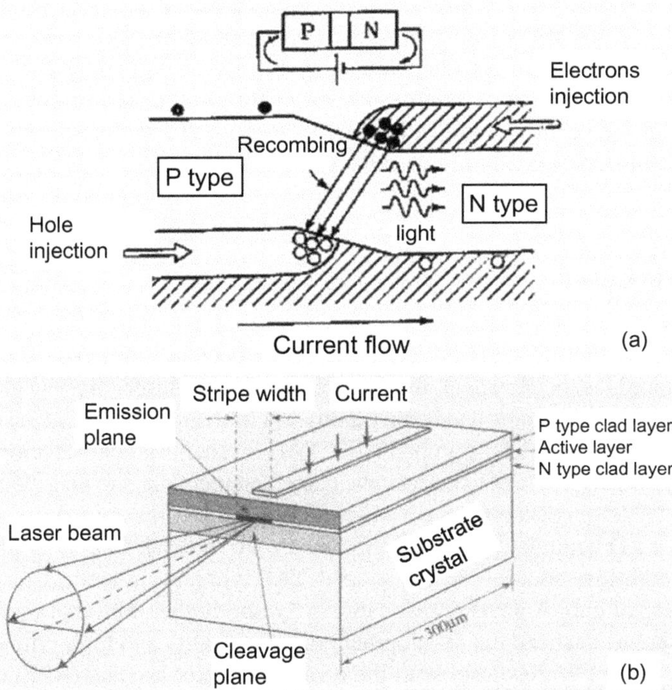

Fig. 1.8 Emission mechanism of diode laser (**a**) and schematic illustration of diode laser device emitting laser beam (**b**)

laser" is generally used, and the name of "LD" is employed as pumping light sources for YAG, disk and fiber lasers.

The laser is proportionally emitted when the current is flown over the threshold. The groups III and V or groups II and VI semiconductors are used, and the wavelength of a diode laser depends upon the composition of the semiconductor and ranges from visible light to infrared light. High-power diode lasers and the lasers for pumping of disk and fiber lasers are developed, and now blue diode lasers are receiving high attention as efficient heat sources for welding of Cu because of higher absorption.

The merits of diode lasers are (1) small, compact, and lightweight, (2) high lasing efficiency (sometimes over 50%), (3) long life, (4) possible power modulation due to the varied currents, (5) superior contribution to the irradiation of high-efficient and high-quality lasers, (6) possible cost reduction due to mass production, and so on. On the other hand, the disadvantages of diode lasers are (1) small laser power from one device, (2) bad focusing properties due to a bad transverse mode, and so on. Recently, these disadvantages have been greatly improved.

The higher power lasers are developed by utilizing the following various measures: (1) improvement of one device in higher power, (2) combining laser beams by fibers, (3) the use of a bar composing of more than ten semiconductor lasers, (4) stacking bars of semiconductors, (5) superimposing laser beams of different wavelengths by polarization beam splitters and dichroic mirrors, and so on.

Diode lasers are used for quenching transformation hardening and cladding of steels, brazing of Zn-coated steels, welding of thin sheets and plastics, and so on. Besides, high-power diode lasers are recently developed to produce deeply penetrated weld beads in steels and aluminum alloys.

1.7 Disk Laser

Lamp-pumped high-power Nd^{3+}:YAG lasers have been used as a typical one of fiber-delivered solid lasers. However, the drawbacks of YAG lasers are low electric-light conversion efficiency and a bad beam quality. As one improvement in the conversion efficiency, laser diode (LD)-pumped Nd^{3+}:YAG lasers have been developed. Thereafter, LD-pumped Yb^{3+}:YAG lasers have been developed, as shown schematically in Fig. 1.9 [11]. A thin disk type of a YAG plate is placed on the coin-sized copper (Cu) plate, and a LD beam is repeatedly irradiated onto the disk-like YAG plate several times. In addition, about 10%Yb^{3+}-doped YAG plate is utilized instead of about 2%Nd^{3+}-doped one. A devised resonant cavity and higher conversion efficiency of 10%Yb^{3+} have enhanced the development of high-quality and high-efficiency, high-power lasers. 16 kW disk laser with the beam parameter product (BPP) of less than 8 mm·mrad can be commercially available and delivered through a fiber of 0.2 mm in diameter.

Fiber-delivered high-power disk lasers of 1.03 μm wavelength are used for welding of car bodies and parts, cutting, brazing, cladding, and so on. And they are used for remote/scanner welding of car parts, welding of thick plates, cutting of thick plates and high-speed cutting of thin steel sheets instead of materials processing with CO_2 laser and YAG laser. Moreover, hybrid welding of thick steel plates with a high-power disk laser and MAG arc is employed for shipbuilding.

Especially, high-power green lasers of 515 nm have recently been developed from high-power disk lasers to use for welding of Cu sheets and plates [6].

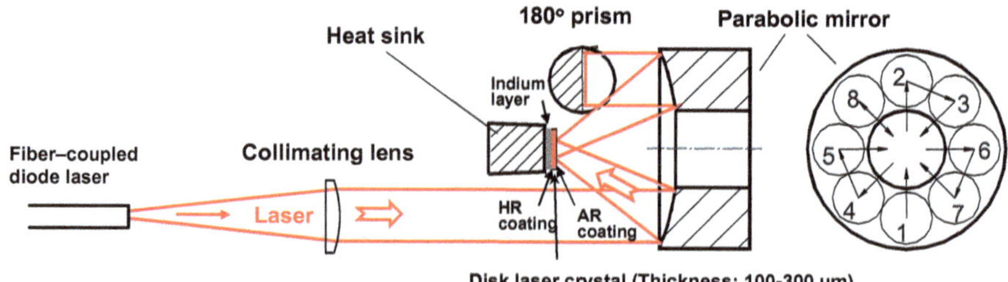

Fig. 1.9 Emission mechanism of disk laser apparatus

1.8 Fiber Laser

Laser is efficiently emitted by LD pumping about 10–20-μm-diameter fiber of high-purity SiO_2 quartz glass doped with Yb^{3+} rare earth element. Such an emitted laser beam or a laser apparatus is called fiber laser. A fiber of double-clad layers is used for fiber laser emission and LD pumping, as shown in Fig. 1.10 [7]. A full reflection mirror and partial reflection mirror are installed in a fiber, as indicated in Fig. 1.11 [7], and thus, the adjustment of a laser beam is not needed, which means easier handling

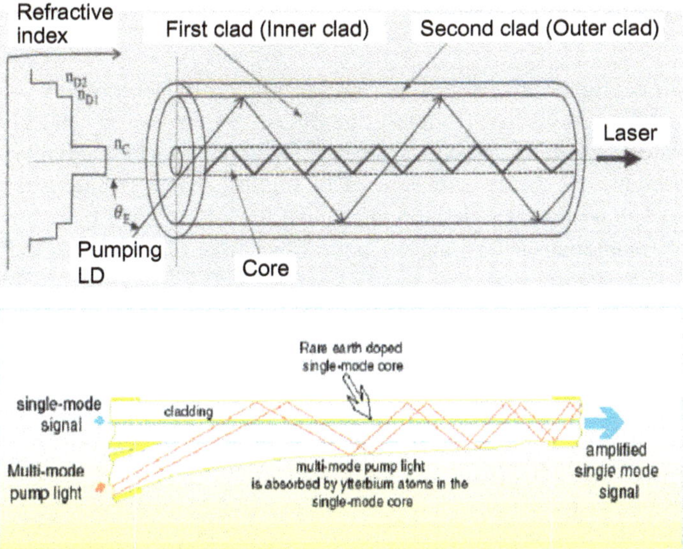

Fig. 1.10 LD pumping and double-clad layers for fiber laser

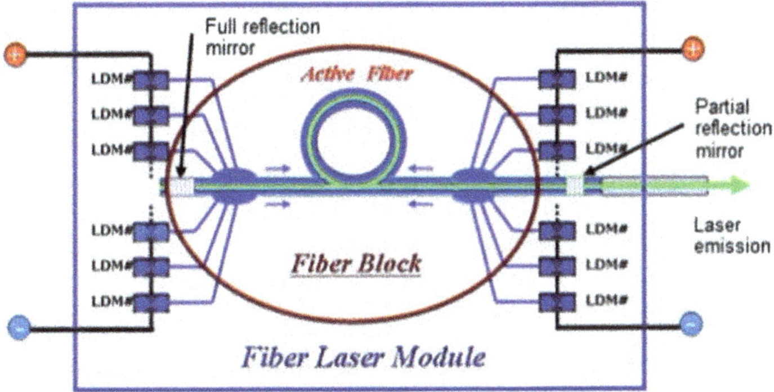

Fig. 1.11 Fiber laser configuration

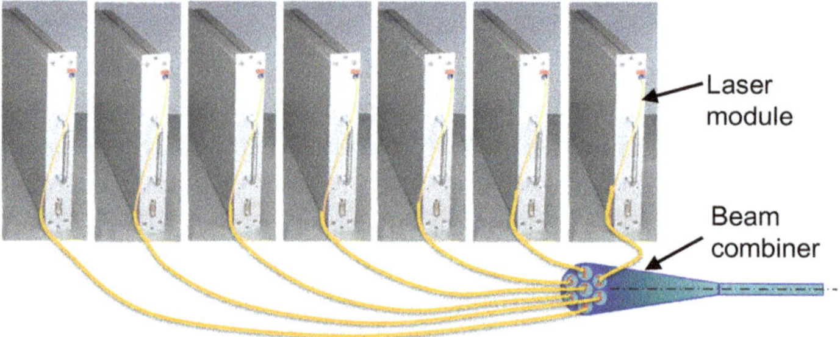

Fig. 1.12 Laser modules and beam combiner for high power of fiber laser

and operation. Higher powers of fiber laser can be easily achieved by using a beam combiner of fibers delivered from laser modules, as shown in Fig. 1.12 [7].

Fiber laser has many advantages such as (1) high beam quality, (2) small and lightweight, (3) easier achievement of high-power, high-intensity, and high-efficiency, (4) long-distance delivery through a fiber owing to 1.07 μm wavelength, and (5) no maintenance. CW single-mode fiber laser of 10 kW in the maximum power and CW multimode fiber laser of 100 or 120 kW in the maximum power are commercially available. 100 kW fiber laser is composed of 90 sets of 1.2 kW laser and is delivered through a fiber of 0.5 mm diameter and 50 m length. On the other hand, the delivery distances of single-mode fiber laser depend upon the laser powers and are generally short, which is the drawback of single-mode fiber laser. The delivery distance of single mode laser is improved these days.

At the present, high-efficiency, high-power fiber lasers are used for welding, cutting, cladding, 3D printing (AM: additive manufacturing), etc. in many industries of cars, railway cars, bridge structures, shipbuilding, and so on. Hybrid welding with fiber laser and MAG arc or CO_2 gas arc is employed in the shipbuilding field in Europe or Japan, respectively.

Low-power lasers of less than 1 kW are used for microprocessing (welding and cutting) of thin sheets, marking, 3D printing, etc. Pulsed fiber lasers of high efficiency are commercially available for the place of pulsed YAG lasers. Ultrashort pulse lasers or green lasers are also developed for microdrilling, microprocessing of Cu sheets, and so on [1].

1.9 Green or Blue Laser

The absorptance of incident light on several metals at room temperature is shown as a function of wavelength in Fig. 1.13 [1, 12]. The absorptance has a tendency of an increase with a decrease in wavelength of the light for any metals. It is understood that the absorptance of Cu is about 50% and 58% for green and blue lasers, respectively.

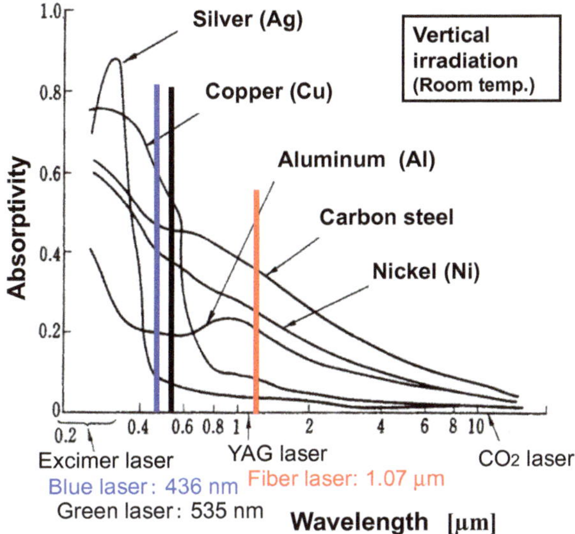

Fig. 1.13 Absorptivity of incident light on several materials at room temperature as function of wavelength, showing higher absorption of blue and green lasers on Cu than YAG, fiber, and CO_2 laser

In other words, green and blue lasers are more advantageous in processing Cu sheets in terms of higher initial absorption than CO_2, YAG, diode, disk and fiber lasers [11].

Green (second harmonic generation; SHG) lasers of 532, 515, and 535 nm in wavelength can be obtained through nonlinear optical crystals such as LBO (LiB_3O_5) or KTP ($KTiOPO_4$) from YAG laser, disk laser, and fiber laser, respectively. The powers of green lasers are normally about half of those of fundamental wave lasers. Nevertheless, about 1–3 kW high power green lasers from disk lasers are developed to weld Cu plates and the stable welding situations are demonstrated [6].

Blue laser is generally one of diode lasers. About 1 kW blue lasers have been developed to melt Cu sheets or to be employed as an effective heat source for additive manufacturing. The drawback of blue laser is that a delivery fiber is not long owing to its shorter wavelength. Presently, the power densities of blue lasers are not so high as to produce deeply penetrated welds in Cu plates. Thus, hybrid welding with blue diode laser and fiber laser or normal diode laser is applied to stably weld Cu plates [13].

1.10 Ultrashort Pulse Picosecond or Femtosecond Laser

Ultrashort or super-short pulse lasers are called picosecond laser (with the pulse width of 1 to 800 × 10^{-12} s) or femtosecond laser (with the pulse width of 1 to 800 × 10^{-15} s) depending upon the pulse widths of lasers shorter than nanosecond (10^{-9} s). The travel distance of a laser beam for 10 femtoseconds is only as short as 0.003 mm (3 μm). Such ultrashort pulsed lasers can be obtained by shortening pulse width due to mode-locking (mode synchronization) technology and by increasing laser energy

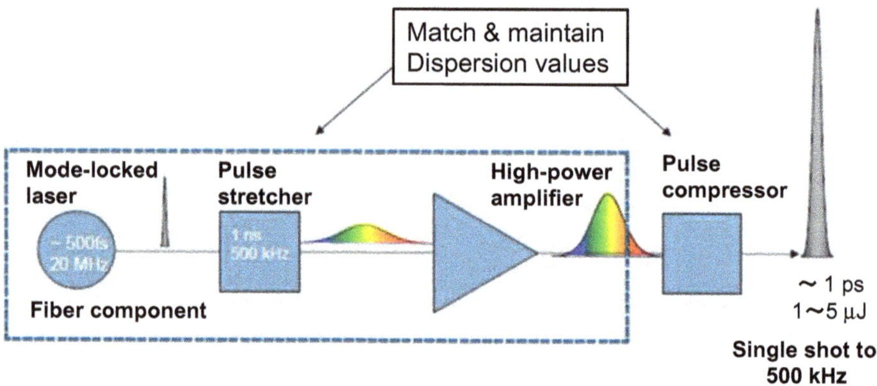

Fig. 1.14 Emission mechanism of ultrashort pulse laser

due to chirped pulse amplification (CPA) technology, as shown in Fig. 1.14 [14]. Mode-locking technology can be realized by appearance of titanium (Ti) sapphire laser, and now disk and fiber lasers can also be utilized. Diffraction grating optical systems are used for the expansion and compression of pulse widths, and are named a pulse stretcher and a pulse compressor, respectively [1].

In microprocessing with nanosecond lasers, debris are formed on the plate surface and microcracks sometimes occur in fusion zones and heat-affected zones (HAZs). However, in processing with picosecond or femtosecond lasers, super-precision or ultraprecision processing is possible by the fact that debris are hardly formed and HAZs are extremely narrow. But it should be known that some debris are generated in processing with a high-power laser. Ultrashort pulsed lasers are applied for precision processing or correction of semiconductors and liquid crystals, processing of transparent materials such as glasses and sapphires, manufacturing of wave guide tubes for optical communication, drilling of engine parts of cars and aircrafts, and so on. Especially, joining or bonding of glasses are actively investigated by employing picosecond or femtosecond lasers [15].

References

1. Katayama S (2019) Very easy book of laser processing. The Nikkan Kogyo Shimbun Ltd. (in Japanese)
2. Katayama S (2009) Laser Welding. J Jpn Weld Soc 78(2):124–138 (in Japanese)
3. Leibinger N (ed) (2007) The laser as a tool. TRUMPF GmbH
4. Sobolev N (1970) Laser and its future. Japanese trans: K. Kobayashi and H. Nakamura, Tokyo-Tosho, Co. (in Japanese)
5. Katayama S (2012) Recent progress in laser welding technology. J Vac Soc Jpn 55(11):471–480 (in Japanese)
6. Brockmann R, Nakamura T (2020) Personal communication. TRUMPF Laser- and Systemtechnik, GmbH. https://www.trumpf.com/

7. Scherbakov E, Kikuchi J (2019) Personal communication. IPG Photonics Corporation. https://www.ipgphotonics.com/
8. Mikame K (2018) Personal communication. TAMARI Industry
9. Nagai H (1989) Laser apparatus for processing. In: Kawasumi H (ed) Cutting-edge laser processing technology, pp 23–45 (in Japanese)
10. Hirata S (2001) Understanding fundamentals and applications of diode laser. CQ Publishing Corp (in Japanese)
11. Nakamura T (2018) The latest laser welding system of TRUMPF. In: Proceedings of the 89th Laser Materials Processing Conference, Osaka, Japan, JLPS, vol 89, pp 55–60 (in Japanese)
12. Miyazaki T, Miyazawa H, Murakawa M, Yoshioka S (1991) Laser processing technology. Sangyo-Tosho (Industry-Library) Ltd., p 18 (in Japanese)
13. Takeda S (2019) Personal communication. Laserline. https://www.laserline.com
14. Laser Concierge Website. https://www.laser-concierge.com (in Japanese)
15. Miyamoto I (2013) Laser welding of glass. In: Katayama S (ed) Handbook of laser welding technologies. Woodhead Publishing, pp 301–331

Chapter 2
Fundamentals of Laser–Materials Interaction and Peripheral Optical System

2.1 Laser Absorption, Transmission, and Reflection of Material

When a laser beam is shot on the material, the absorption, reflection, transmission, refraction, and scattering occur, as shown in Fig. 2.1 [1]. These degrees depend upon the kinds and sizes of the materials such as metals, alloys, ceramics, plastics, glasses, and so on. The laser absorptivity or absorptance (absorption coefficient or coupling coefficient) of the plate material, A, is expressed as $A = 1 - R - T$, where R and T are reflectivity (optical reflectance) and transmittance, respectively. The laser absorptance of metals and alloys, A, is expressed as $A = 1 - R$, since the transmittance is almost zero ($T = 0$). In addition, Rayleigh scattering of a laser beam occurs when it is shot on ultrafine particles if the wavelength of the laser beam, λ, is larger than the sizes of particles, d ($\lambda \gg d$).

The metals have valence electrons which can move around freely all over the crystals, and therefore, the conductivity of electricity is high. The laser absorptance of the metal is roughly proportional to the conductivity. When a laser beam is irradiated on the metal, the laser is hardly absorbed because of easily movable free electrons. However, partly, intraband transition of free electrons occurs, and free electrons also interact phonons due to lattice vibration, characteristic vibration due to dipoles, and potential disorders due to imperfection, defects, and impurities of crystals [2]. Accordingly, part of laser energy is absorbed into the metal and the absorbed energy is transduced to thermal energy, and consequently, the temperature of the metal rises. The metal is melted, and a molten pool is formed. Finally, when the laser power density is high in welding, evaporation occurs to produce a cavity or a keyhole in the molten pool.

The absorptivity of several metals is already shown as a function of the light wavelength in Fig. 1.13, where the light is irradiated vertically on the metallic plates at room temperature. The absorptivity generally increases with a decrease in the wavelength. A coupling (absorption) coefficient of laser energy into the metal or the

S. Katayama, *Fundamentals and Details of Laser Welding*,
Topics in Mining, Metallurgy and Materials Engineering,
https://doi.org/10.1007/978-981-15-7933-2_2

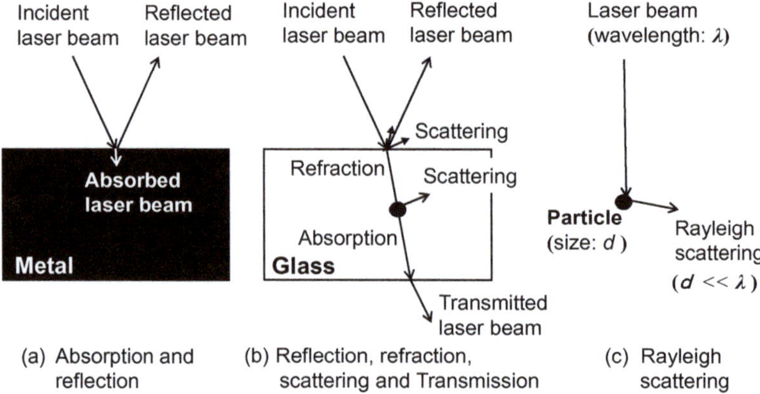

Fig. 2.1 Schematic representation of physical interaction of absorption, reflection, transmission, refraction, and scattering between laser beam and material

laser absorptivity of a metal is schematically shown as a function of temperature in Fig. 2.2 [1, 3]. The absorptivity generally increases slightly in the solid state as the temperature rises, and oxidation and/or strain-induced forming can increase the absorption of a laser beam. The absorptivity increases appreciably in the molten state when the metal melts and largely in the condition of keyhole formation because the laser beam reaches the bottom of a keyhole due to multi-reflection.

The laser absorptivity (absorption) of Type 304 stainless steel and A5052 aluminum alloy was measured during actual laser welding by using the calorimetry as schematically shown in Fig. 2.3 [4]. The beam quality, spot diameter, and focusing situations are different between YAG laser and fiber laser. The spot size of a YAG laser beam is about 0.6 mm and typical. On the other hand, a fiber laser can be

Fig. 2.2 Effect of surface temperature on laser coupling coefficient or absorptivity of material

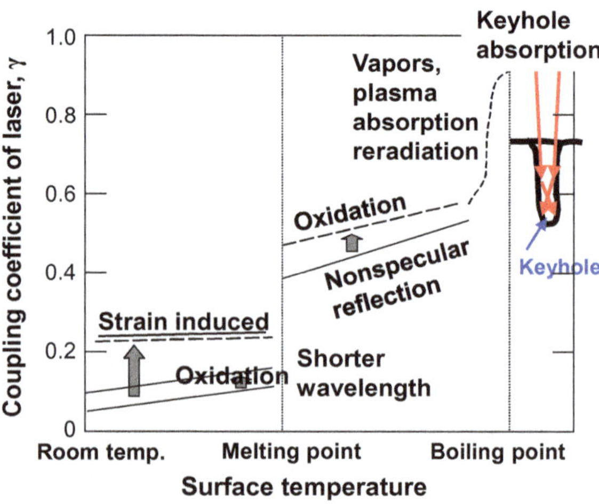

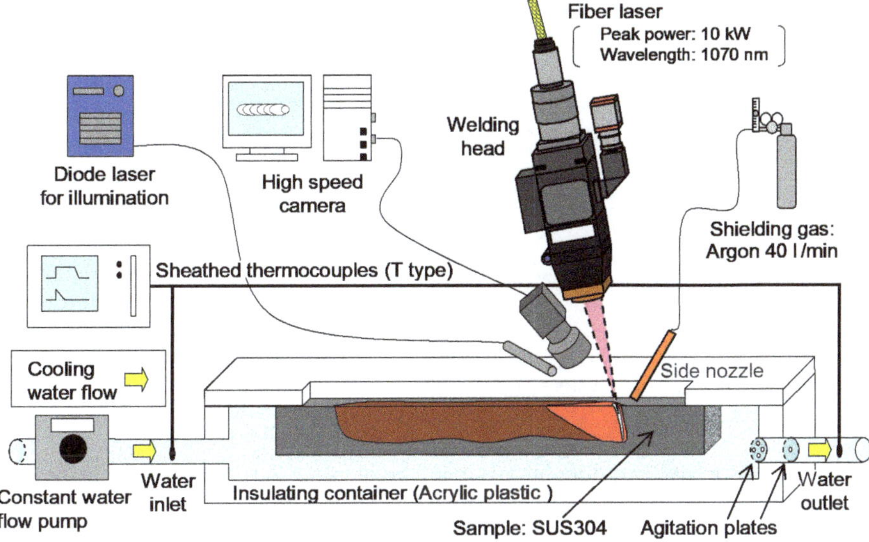

Fig. 2.3 Schematic illustration of calorimetry method for measurement of laser absorptance of material

focused on a smaller spot size. The absorptivity of Type 304 steel obtained during welding with CW fiber laser or CW YAG laser under various defocused conditions is compared together with the laser focusing situations and the weld bead geometries in Fig. 2.4 [5]. Under the same conditions of 2.5 kW power and 1 m/min welding

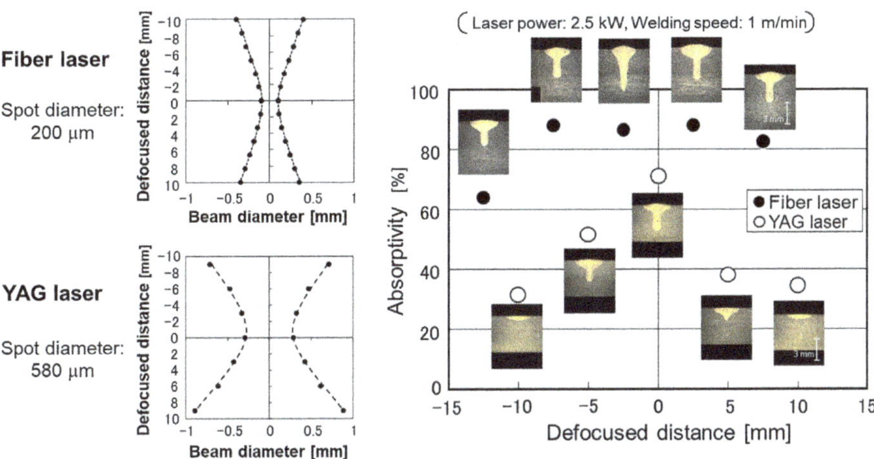

Fig. 2.4 Comparison of beam focusing situations, cross-sectional weld beads made in Type 304 plates at different defocused distances and their corresponding laser absorptivity during welding between CW fiber laser and CW YAG laser

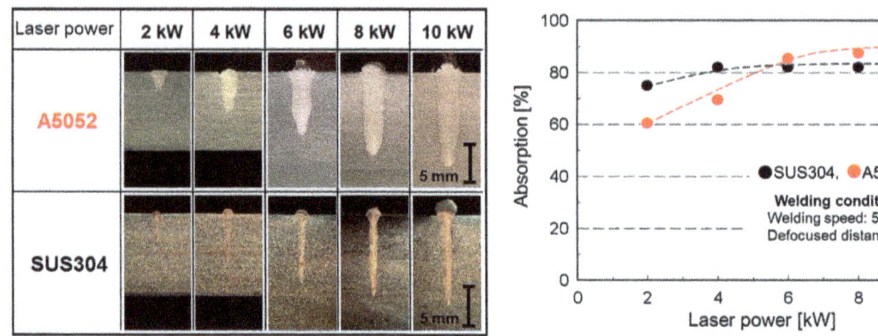

Fig. 2.5 Comparison of cross-sectional fiber laser weld beads produced in aluminum alloy A5052 and austenitic stainless steel Type 304 at different powers, and effect of laser power on laser absorption during welding of A5052 and Type 304

speed, the absorptivity of fiber laser is much higher than that of YAG laser at the same defocused distances although in both cases the absorptivity generally decreases with an increase in the defocused distance. This is attributed to the reason why the keyhole is easily formed due to the smaller beam spot diameter and the consequent higher laser power density of the fiber laser than the diameter and the density of YAG laser. The effect of wavelength should be small because the laser absorptivity decreases under the defocused conditions. The effects of laser power and welding speed on the absorptivity were investigated during welding of Type 304 steel and A5052 alloy with the fiber laser of 0.2 mm in spot diameter. The effect of laser power on the absorptivity is compared between Type 304 steel and A 5052 alloy, as indicated in Fig. 2.5 [6]. The laser absorptance of stainless steel is higher at low powers, while that of aluminum alloy is higher at high powers of more than 6 kW. The increase in the laser absorption with an increase in the power and higher absorption of Al alloy at higher powers are attributed to the formation of a larger size of a keyhole inlet.

The effect of welding speed on the laser absorption together with a keyhole and a molten pool observed during laser welding is exhibited in Fig. 2.6 [4, 6]. The absorptivity decreases with an increase in the welding speed. As the welding speed increases, the molten pool becomes smaller, and the melt zone in front of a keyhole is also narrower (or thinner). That is, at higher welding speeds, a part of a laser beam is irradiated on the thin melt zone and solid metal surface; consequently, the laser reflection or reflectivity increases, and the ratio of a laser beam absorbed into a keyhole decreases. This signifies that the laser absorptance of solid-state metal surface is extremely low while that of a keyhole is rather high.

In the case of glass or plastics, the transmissivity or light transmittance of a laser beam varies with laser wavelength, and thus the absorptivity and transmissivity should be measured and evaluated for each laser beam. For example, excimer lasers of short wavelengths and CO_2 laser of long wavelength are absorbed in glasses. Namely, processing of glasses is feasible with lasers of short or long wavelength.

In processing materials with a polarized laser beam, the effect of polarization on the absorption or reflection of the materials should be considered.

Welding speed	2 m/min	3 m/min	6 m/min	10 m/min	15 m/min
High speed images					1 mm
Schematic illustration	Molten pool Keyhole Laser spot				
Beam ratio in keyhole	100 %	100 %	85 %	69 %	66 %
Absorption	84 %	84 %	79 %	74 %	68 %

(Laser power: 10 kW, Laser spot diameter: 200 μm, Shielding gas: Argon 40 l/min)

Fig. 2.6 High-speed video observation photographs of keyhole in molten pool during laser welding at different speeds and their sketches, and effect of welding speed on ratio of laser beam in keyhole and laser absorption measured during welding of Type 304 steel

2.2 Effect of Polarization on Laser Absorption of Material

The polarization of a laser beam exerts a considerable effect on the quality of the material subjected to welding, cutting, drilling, or surface quenching transformation hardening.

Laser is an electromagnetic wave with electric field and magnetic field. In materials processing, the laser is representatively treated as electric field because it has a greater interaction with electrons in the material. The incident plane of a laser beam is defined as a plane including the laser incident direction axis and the line perpendicular to the material surface, as shown in Fig. 2.7 [2]. The absorption or reflection depends upon the incident angle of a laser beam, and the laser reflectivity

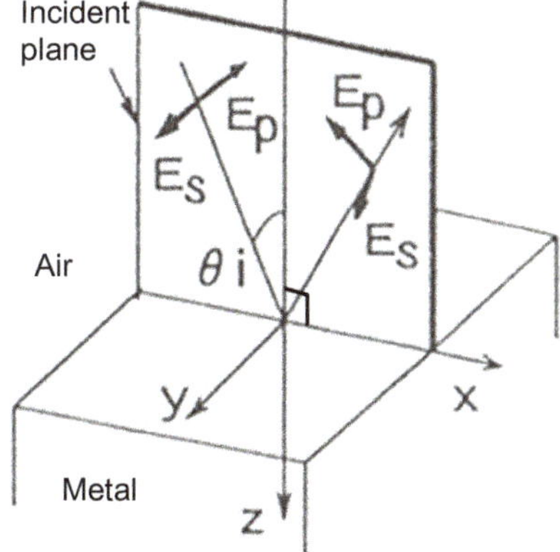

Fig. 2.7 Definition of incident plane and light polarization (E_s and E_p) to metal

of p-polarization and s-polarization to the metallic material is varied as a function of the incident angle. An example of the effect of polarization on the reflectivity of stainless steel for CO_2 laser of about 10 μm wavelength is shown at the temperature of 300, 1000, and 3000 K in Fig. 2.8 [7]. The rise in the temperature decreases the reflectivity slightly. In the case of s-polarization, the reflectivity increases gradually with an increase in the angle of incidence. On the other hand, *p*-polarization, the reflectivity decreases extremely at the angle of about 85°.

It has been reported that, when the steel plates are subjected to welding with polarized and non-polarized CO_2 lasers, the penetration depths of weld beads are different at high welding speeds. The results are shown in Fig. 2.9 [7]. In the case of low welding speeds, almost all energy of a laser beam is incorporated into a keyhole, and therefore, the penetration depths of three kinds of weld beads are nearly equal. In

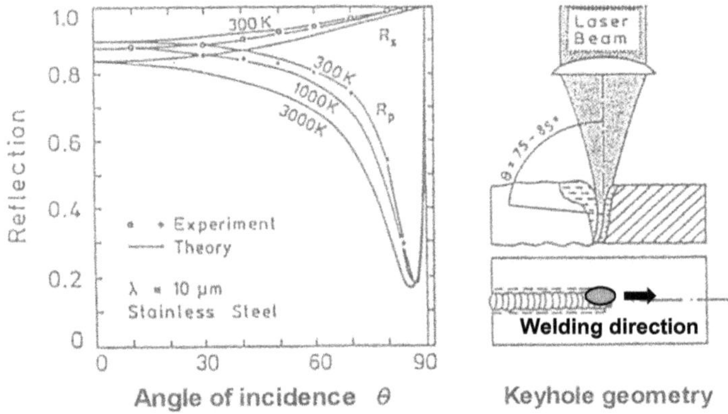

Fig. 2.8 Reflection of *s*- and *p*-polarized laser against steel as function of incident laser angle, and consideration of relationship between laser beam and keyhole wall during laser welding

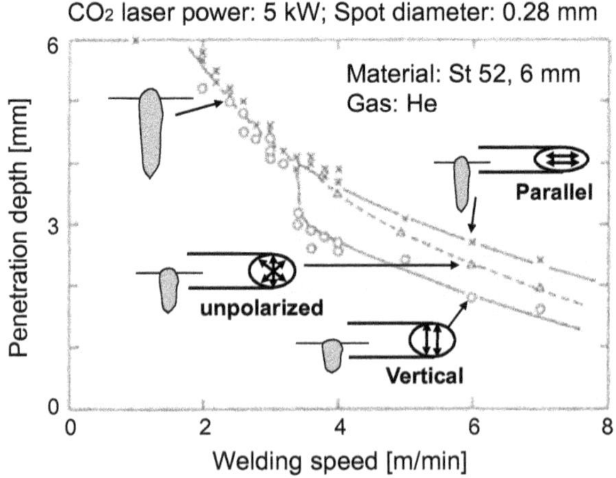

Fig. 2.9 Effects of polarization and welding speed on penetration depths of CO_2 laser weld beads in steel

the case of higher welding speed, it is more difficult to melt the front area of a keyhole and a laser beam should be irradiated on the front wall of a keyhole. Consequently, the reflectivity of an s-polarized laser beam is higher at a higher incident angle, and the keyhole depth becomes shallower, resulting in a shallower penetration weld. On the other hand, a p-polarized laser beam should be sufficiently absorbed along the front wall of a keyhole because of extremely low reflectivity at the high incident angle. Consequently, deeper welds are obtained at high welding speeds. The penetration depths of welds produced with a circular or random (unpolarized) laser beam are middle between those made with s- and p-polarized laser beams.

2.3 Effect of Laser-Induced Plume or Plasma on Laser Propagation, Reflection, and Absorption

When a laser beam is irradiated on the metal sheet or plate, part of the laser energy is absorbed due to the interaction of the laser with the electrons in the laser-irradiated area of the metal, and the temperature of the area rises, melting takes place and evaporation occurs from the laser irradiation part, as schematically shown in Fig. 2.10 [1, 8]. If the recoil pressure due to the evaporation is high, the surface of a molten pool is depressed, and finally a deep cavity called a keyhole is formed. Evaporated atoms and clusters, leading to the formation of ultrafine particles or fumes over the metal plate, are ejected upwards from the keyhole inlet, and the melts in the keyhole wall or around the keyhole inlet are sometimes spouted out as spatters or molten droplets.

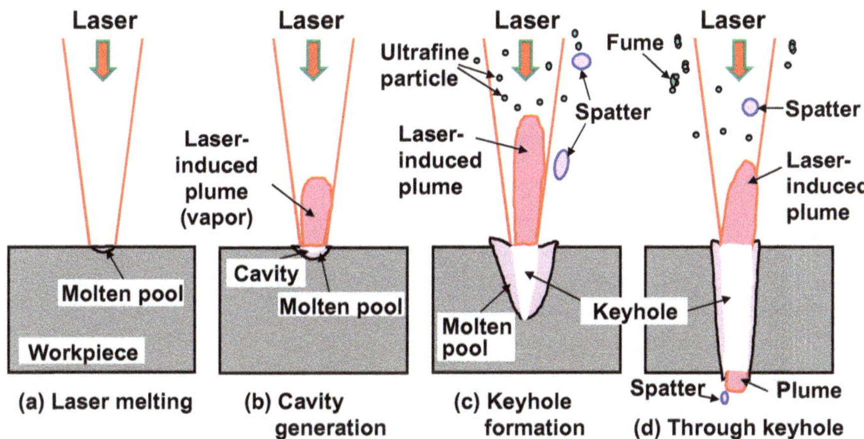

Fig. 2.10 Schematic illustration of physical phenomena of melting, molten pool formation, keyhole evolution, plume ejection, spattering, formation of ultrafine particles and fume formation during laser irradiation

The evaporated substances or matters, consisting chiefly of neutral metal atoms, are extremely bright fluorescent substances at high temperatures and are called as a laser-induced plume from the shape similar to the plume of a bird. Besides, if the bright substances are ionized into ions and electrons, it is called a laser-induced plasma or a weekly ionized plasma [9]. The evaporated atoms produce clusters (of several nm in size) and ultrafine particles (of about 10–200 nm in size) of metal atoms or oxides by bonding due to their collision [9]. Finally, floating fumes are seen over the workpiece.

Subsequently, the formation situation of a laser-induced plume and/or plasma can be discussed from the results of high-speed video observation, as shown in Fig. 2.11 [8]. In CO_2 laser welding in argon (Ar) or nitrogen (N_2) shielding gas, Ar or N plasma may be formed over the workpiece in addition to laser-induced plume ejected from the keyhole intel at the power levels of 5 kW. The gas plasma renders the weld penetration shallower to a slight degree. If the power is increased to more than 10 kW, larger sizes of Ar or N plasmas are periodically formed over the workpiece, resulting in the formation of by far shallower welds [10]. On the other hand, in welding with high-power CO_2 laser in helium (He) shielding gas, gas plasmas are not formed because of high ionization potential and therefore deeply penetrated weld beads are produced. It is interpreted in high-power CO_2 laser welding in Ar or N_2 shielding gas by considering that the laser energy is absorbed by the inverse Bremsstrahlung of gas plasma (where the laser is absorbed by electrons moving in the electric field of ions (atomic nucleus) and then the moving directions of electrons are bent or travelled

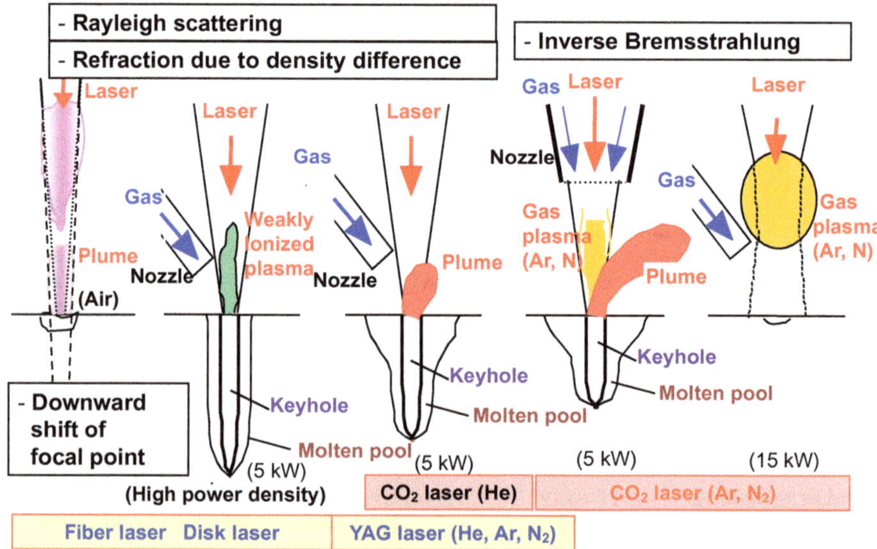

Fig. 2.11 Schematic representation of typical physical phenomena during welding with practical lasers such as fiber, disk, YAG and CO_2 laser, showing plume evolution from keyhole and partly plasma formation, and their effects on weld penetration

with increasing speeds) [11, 12]. The active degree of inverse Bremsstrahlung is proportional to the square of laser wavelength, and the effective levels of YAG, disk, and fiber laser are about 100 times smaller than those of CO_2 laser. Gas plasmas are absent in welding with lasers of about 1 μm wavelength in Ar or N_2 shielding gas. No gas plasmas are formed in He shielding gas.

In welding with the lasers of less than 1.5 μm in wavelength, the Rayleigh scattering should occur due to the formation of clusters and ultrafine particles [9]. The Rayleigh scattering is inversely proportional to the 4th power of the laser wavelength. The lasers of shorter wavelengths exert a greater effect, and the effects of YAG, fiber, and disk lasers of about 1 μm wavelength are rather small. If a plume is not suppressed and becomes tall in welding with any lasers, a long range or a wide area of small density and small refraction index is formed above the specimen due to high temperatures. It induces the welding phenomena that a laser beam is refracted and/or the focal point of the laser beam shifts downwards [13]. Consequently, the laser power density at the workpiece surface is reduced to a greater degree, and accordingly, the weld penetration depth is drastically shallower by changing from keyhole mode to heat-conduction mode of welding. The suppression of plume height by using a shielding gas, air blower, fan, and so on is necessarily important at all times to stably produce a deep laser weld bead.

2.4 Focusing Optics, and Measurement and Monitoring of Laser Power Density

Laser welding is performed at high power densities of a small-diameter beam focused by a lens or a mirror. The welding results of joint properties such as penetration depths, bead widths and joining areas, defects sensitivity such as cracking susceptibility, spattering and porosity formation, or mechanical properties such as strength, elongation, and toughness are affected by laser welding conditions, especially, laser power, intensity (power density) and traverse electromagnetic (TEM) mode. It is important to know the power, mode, and focusing properties of a laser beam in obtaining satisfactorily sound welds. The focusing situation can be chiefly determined by the optics employed. The small beam diameter of a laser can be obtained by the following procedures: (1) the use of a focusing optics of shorter focal length, (2) the expansion of a laser beam incident on the focusing lens due to the utilization of a beam expander, (3) the use of a laser with a high beam quality, and (4) the use of a laser with a shorter wavelength.

A single Gaussian mode laser is a fundamental mode of an ideal beam because it can be focused in the smallest spot. Focusing situation of an actual multi-mode laser beam is evaluated by comparing that of a single Gaussian mode laser, as shown in Fig. 2.12 [1, 14]. A laser beam is usually in a multi-mode, and the beam quality is conventionally assessed by the factor M^2 (called M square) or K value ($=1/M^2$), in which M^2 is expressed as a beam radius ratio of a multi-mode to a single Gaussian

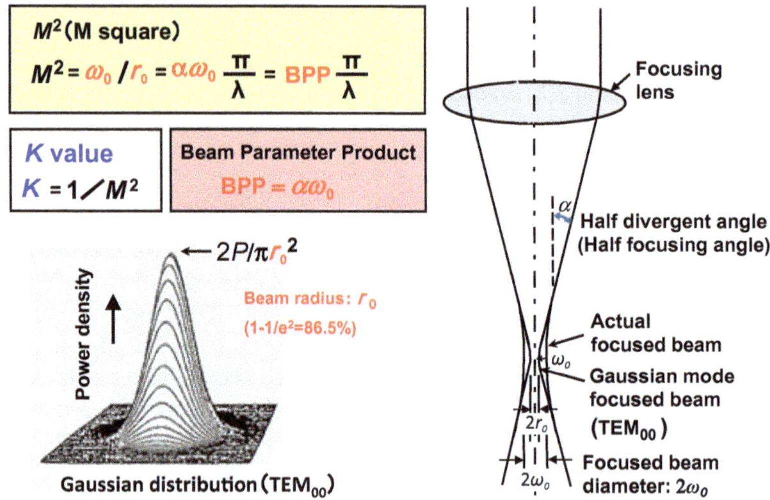

Fig. 2.12 Definition of beam quality evaluation factors, and comparison of focusing situations between actual mode laser and Gaussian mode laser

mode, and ω_0/r_0 or $\alpha\omega_0\pi/\lambda$, where ω_0 is a radius of a focused multi-mode laser beam, r_0 is a radius of a focused single Gaussian mode laser beam, α is half of a divergent angle or a focusing angle, and λ is the wavelength of a laser beam. It is evaluated that the beam quality is better as the value of M^2 or K approaches to 1 (one).

In the place of M^2 or K, recently, the value of beam parameter product (BPP) expressed as $\alpha\omega_0$ is employed in the unit of mm·mrad because it does not include the effect of laser wavelength. It is judged that the beam quality is better as the value of BPP is smaller. A high-quality laser has three following advantages [15]. (1) In the case of the lens with the same focal length, the laser can be focused into a smaller spot diameter, achieving a higher power density leading to deeper penetration or higher welding speed. (2) In the case of the same focal length and the same spot diameter, smaller-sized lightweight focusing optics can be used, resulting in easier handling of a robot. (3) In the case of the use of the same spot diameter, the focusing optics of a longer focal length can be used, leading to improvement in workability. Focusing situation can be compared among typical lasers with different BPPs, as shown in Fig. 2.13 [16]. It is understood for a fiber laser of the smallest beam size that the lens of the smallest diameter or the longest focal length can be used by the emission of a high-quality beam. Namely, fiber lasers can extend the working distance, leading to the improvement in workability.

The laser powers, power densities (intensities), beam modes, and focusing situations are extremely important parameters in laser welding, and thus, these parameters are measured or monitored. At present, the apparatuses, which can measure these parameters, are commercially available except for extremely high power or high power density. An example of one apparatus and its schematic presentation and one

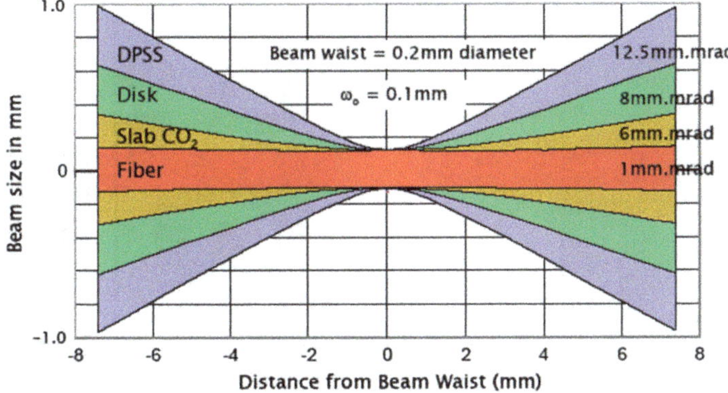

Fig. 2.13 Comparison of focusing situations among industrial lasers of different BPP values

example of measurement results of power density and focusing situation are shown in Figs. 2.14 and 2.15, respectively [17]. Power density is measured by detecting a laser energy into a pinhole probe rotating and traveling at small steps during laser shooting, and focusing situation is obtained by measuring power profiles at different setting locations. Moreover, an apparatus exhibiting focusing situation directly by displaying interaction of a laser beam to a certain low-density gas environment is developed and is also commercially available, as shown in Fig. 2.16 [18]. Besides, various simple measurement methods using an acrylic plate or a carbon sheet are suggested to know intensity (power density) distribution, a beam diameter effective to processing and a focal point of a laser beam by displaying evaporated pattern traces after laser irradiation. Special apparatuses, in which laser power density may also be displayed on the monitor by using a laser beam after a great reduction in transmitted

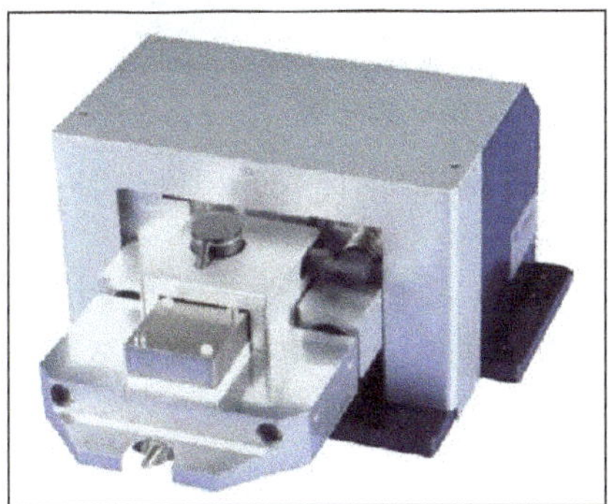

(a) Monitoring device

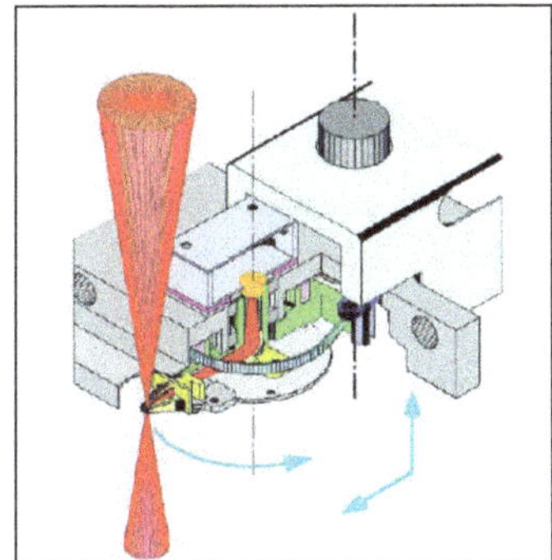
(b) Schematic monitoring device

Fig. 2.14 Monitoring apparatus for measuring laser power density, beam diameters, focusing situation and beam quality evaluation factors such as BPP and its schematic presentation

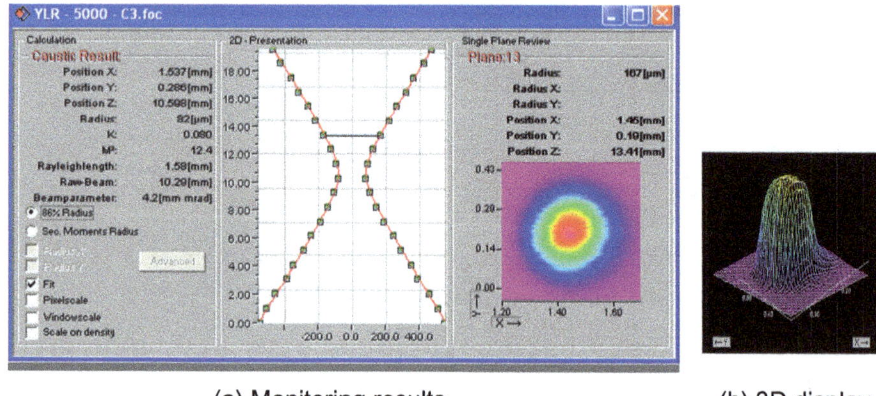

(a) Monitoring results (b) 3D display

Fig. 2.15 Examples of measurement results of power density, BPP, beam mode, and focusing situation

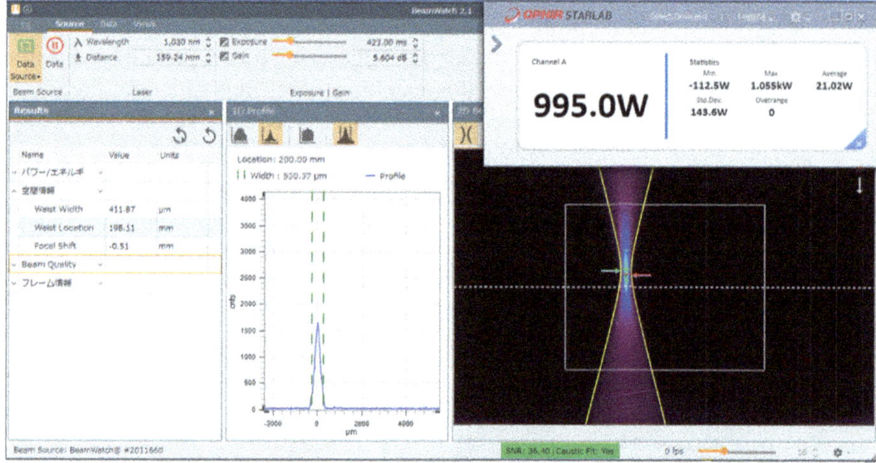

Fig. 2.16 Focusing situation measured directly from interaction of focused laser beam and certain gas component

energy, are commercially available. These are used for visualizing power densities of low-power CW lasers or pulsed lasers.

2.5 Fiber Delivery System

Although the beam delivery of a CO_2 laser is difficult, normal solid-state lasers of about 1 μm in wavelength such as a YAG laser, a disk laser, a fiber laser, and

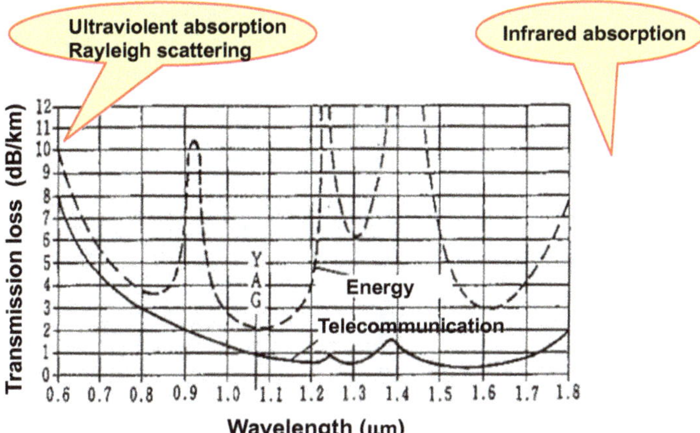

Fig. 2.17 Transmission losses of fiber for delivery of energy or telecommunication as function of wavelength

a diode laser can be delivered from the apparatus to the working place through fine silica (SiO_2) fibers. The transmission losses of fiber for delivery of energy or telecommunication are shown as a function of wavelength in Fig. 2.17 [19]. Silica fibers can easily transmit the laser of about 1.06 μm, while the short wavelength lasers and long wavelength lasers reduce transmissivity due to ultraviolent absorption or Rayleigh scattering and infrared absorption, respectively. The laser beam of about 1 μm in wavelength can travel about several meters to 200 m without any appreciable loss of power or energy through the fiber core of typically about 100–600 μm in diameter. These silica fibers are the heart of fiber optic laser cables. The fiber optic laser cables consist of the core, cladding, and protective plastic coating. The core diameter depends upon the power and quality of a laser beam transported inside. The cladding surrounds the fiber core and ensures the laser beam inside the core. The protective plastic coating forms a durable and robust shell for the brittle glass fibers and makes it possible to bend the glass fibers. A metallic sheet tube is also used to protect a fiber optic laser cable.

There are three types in fibers: step-index (SI type) fibers, graded-index (GI type) fibers, and single-mode (SM type) fibers, as shown in Fig. 2.18 [1]. SI-type fibers of about 50 μm to 1 mm in diameter are generally used for high-power CW lasers of more than 500 W power, and GI-type fibers of about 0.2–1 mm in diameter are normally employed for pulsed (YAG) lasers of less than 500 W power. GI fiber of Ge doping can keep high beam quality but damages easily at high laser powers. SM-type fibers, which are of less than 20 μm in core diameter, can deliver a fundamental mode of a laser beam of several kW powers. The distance capable of laser beam delivery is shorter as the laser power is higher. As seen in Fig. 2.17, the green and blue lasers of shorter wavelengths have difficulty in delivering their beams through normal silica fibers. New fibers have been developed for high-power green and blue lasers [20].

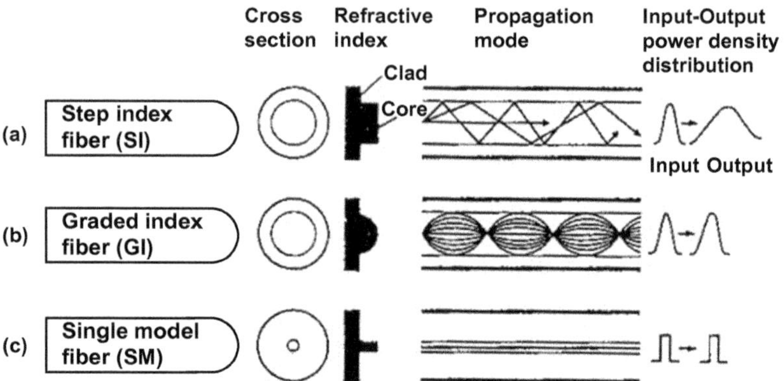

Fig. 2.18 Comparison of various optical features between step-index (SI type) fibers, graded-index (GI type) fibers, and single-mode (SM type) fibers, showing characteristic differences in cross section of fiber, refractive index, propagation mode and input–output power density distribution

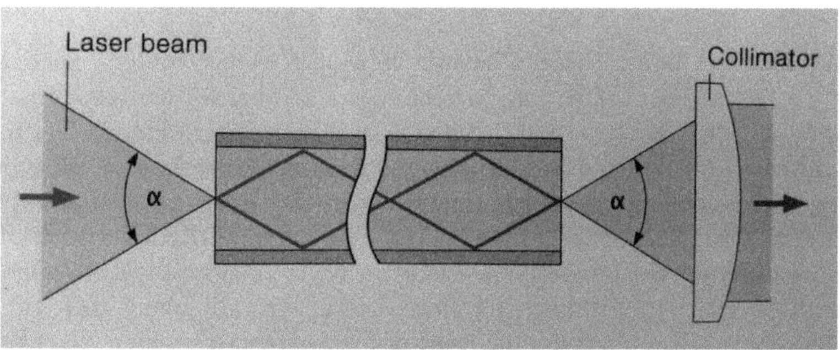

Fig. 2.19 Laser beam entering and exiting fiber at same angle

It is interesting to know that the laser beam exits the fiber at the same angle at which it has been coupled into the fiber, as exhibited in Fig. 2.19 [21]. After the beam leaves the fiber, it is normally collimated.

A laser beam entering the fiber is guided only if it is focused within the acceptance angle of the fiber, as shown in Fig. 2.20 [21]. If the angle of the coupled laser beam is larger than the acceptance angle, the cladding and the plastic coating can be damaged. It is therefore noted that the angle of the coupled laser beam must always be smaller than the acceptance angle. A fiber optic laser cable can guide a laser beam even through curves owing to their flexibility. It should be also noted that the laser beam escapes from the core and gives damages to the cable if the cable is bent at the small radius. It is better to keep the bent radius of the cable larger.

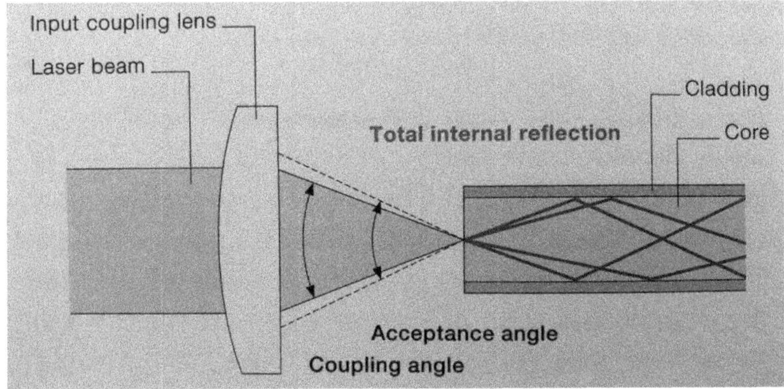

Fig. 2.20 Coupling angle into fiber smaller than acceptance angle

References

1. Katayama S (2019) Very easy book of laser processing. The Nikkan Kogyo Shimbun Ltd. (in Japanese)
2. Kudou M (1996) Fundamental of optical physics. Ohmsha Ltd. (in Japanese)
3. Roessler DR (1986) The industrial annual handbook, 1986 Edition. Penn Well Book, pp 16–30
4. Kawahito Y, Matsumoto N, Abe Y, Katayama S (2009) J Jpn Weld Soc 27(3):183–188 (in Japanese)
5. Katayama S, Kawahito Y, Mizutani M (2012) Phys Proc 39:8–16
6. Kawahito Y, Matsumoto N, Abe Y, Katayama S (2011) J Mater Process Technol 211:1563–1568
7. Beyer E, Behler K, Herziger G (1988) In: Proceedings 5th international conference lasers in manufacturing (LIM-5), pp 233–240
8. Katayama S (2009) Laser welding. J Jpn Weld Soc 78(2):124–138 (in Japanese)
9. Matsunawa A, Yoshida H, Katayama S (1984) Proc ICALEO '84 44:35–42
10. Seto N, Katayama S, Matsunawa A (2000) J Laser Appl 12(6):245–250
11. Miyamoto I (1991) In: Proceedings 26th laser materials processing conference, JLPS, pp 1–17 (in Japanese)
12. Tsukamoto S (2001) J Jpn Laser Mater Process. JLPS 8(1):37–41 (in Japanese)
13. Oiwa S, Kawahito Y, Mizutani M, Katayama S (2011) J Laser Appl 23(2):022007-1–7
14. Arai T, Kutsuna M, Miyamoto I (1993) Fundamental of laser processing (last Volume). Machinist Publishing Co.
15. Bachmann F, (2004) In: Proceedings of the 61st Laser Materials Processing Conference, Osaka, Japan, JLPS, vol 61, pp 16–29
16. O'neil W, Sparks M, Vamham M, Horley R, Birch M, Woods S, Harker A (2004) In: Proceedings of the 23rd ICALEO, LIA, Fiber & Disc Laser Session, pp 1–7 (CD)
17. PRIMES Website (Power Monitor) https://www.primes.de./en/
18. Ophir (Beam Watch) https://www.ophiropt.com/laser-measurement
19. Nagai H (1989) Laser apparatus for processing. In: Kawasumi H (ed) Cutting-edge laser processing technology, pp 23–45 (in Japanese)
20. Brockmann R, Nakamura T (2020) Personal communication, TRUMPF Laser- and Systemtechnik, GmbH. https://www.trumpf.com/
21. Leibinger N (ed) The laser as a tool. Trumpf GmbH

Chapter 3
Fundamentals and Features of Laser Welding

3.1 Fundamentals of Laser Welding

Laser is a high power density and high-energy density heat source, and therefore, laser welding is recognized as advanced materials joining because it is possible to perform high-speed welding, deep penetration welding, or high-precision welding.

Representative power density profiles of laser, electron beam, plasma and arc, and the consequent weld bead geometries are schematically compared in Fig. 3.1 [1, 2]. The power density of a laser beam is substantially high and equivalent to that of an electron beam and is much higher than that of arc or plasma. Consequently, a deep, narrow keyhole can be formed during welding with a high power density beam of laser or electron, and a deep, narrow weld can be effectively produced. An arc can produce a shallow cavity at high current but no keyhole, and a plasma can sometimes form a shallow keyhole. As a result, those welds are shallower than laser welds. Moreover, the traveling speed of laser welding is higher than that of arc or plasma welding. In a schematic presentation, the penetration of an electron beam weld is deeper than that of a laser beam at a low welding speed. This is attributed to the environment difference that electron beam welding is always performed under the vacuum conditions although laser welding is normally carried out in a shielding gas at 1 atm. Sound laser welds of deeper penetration equivalent to electron beam welds can be produced under the vacuum conditions.

There are various kinds of laser-welded joints and seams, as shown in Fig. 3.2 [3]. Generally, butt, lap, or fillet welding of butt joint, lap joint, or T joint is often used. Linear, annular circled (circumferential) and circumferential axis welding are also employed. Seam welds are formed by welding continuously with CW (continuous wave) laser or by repeating spot welds with pulsed wave (PW) laser.

S. Katayama, *Fundamentals and Details of Laser Welding*,
Topics in Mining, Metallurgy and Materials Engineering,
https://doi.org/10.1007/978-981-15-7933-2_3

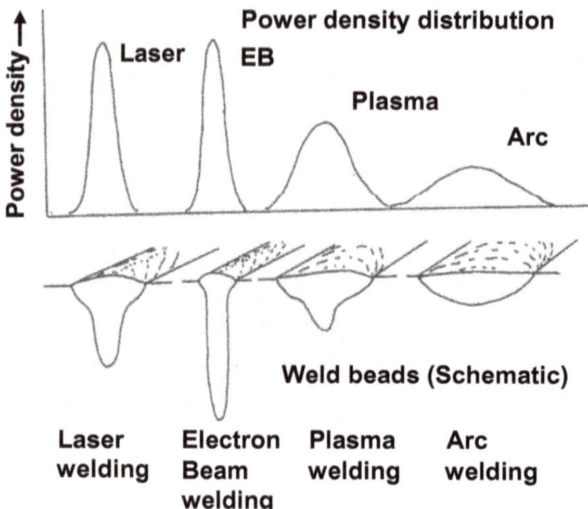

Fig. 3.1 Representative power density profiles of laser, electron beam, plasma and arc, and schematic illustration of their corresponding weld bead geometries

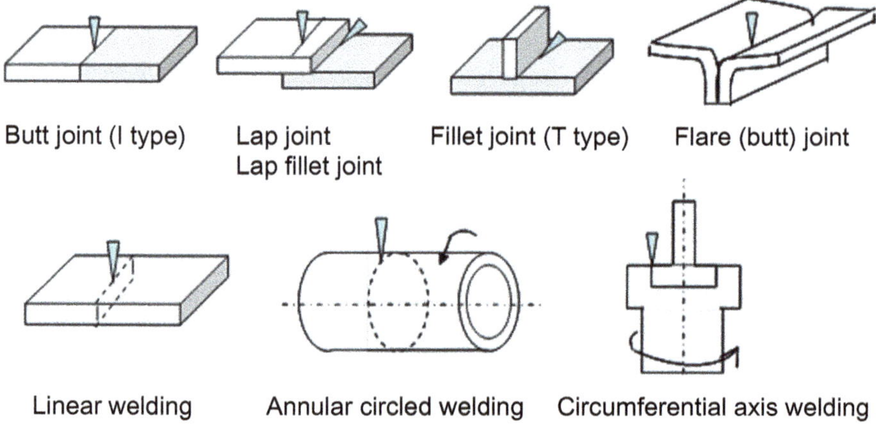

Fig. 3.2 Various kinds of laser-welded joints called butt, lap, or fillet welding of butt joint, lap joint and T joint, and linear, annular circled, and circumferential axis welding

Typical phenomena during PW or CW laser welding are schematically illustrated in Fig. 3.3 [2, 3]. Depending upon the irradiation time and the laser power density, spot or bead welds are produced in the morphology of a heat-conduction type or a keyhole type, respectively. In the case of low laser power density, the spot welds and the bead welds are shallow, and generally less than 0.6 mm deep and less than 2 mm deep, respectively. In the case of high laser power density, a keyhole is formed to produce a deeply penetrated weld. A keyhole type of spot welds with PW laser

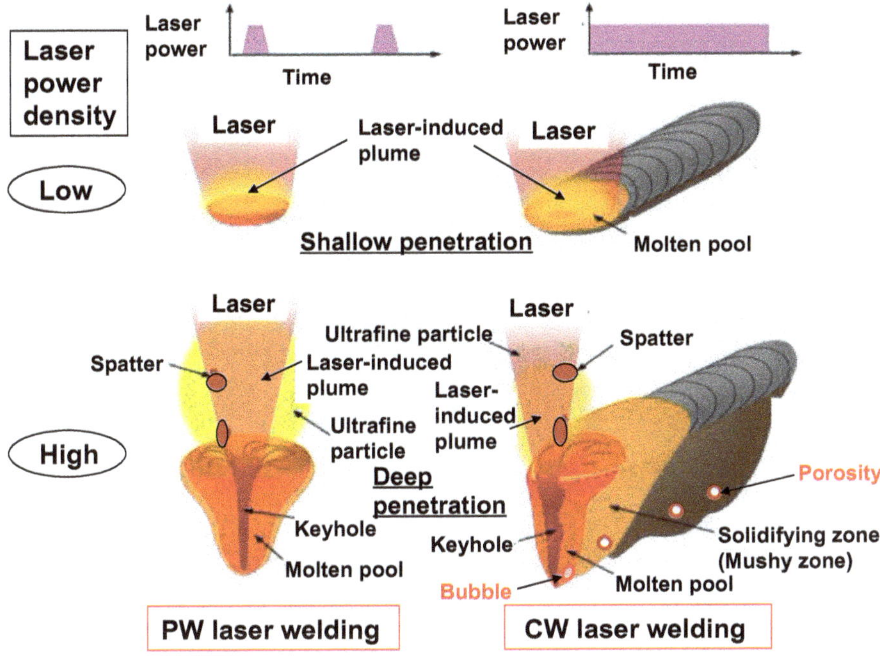

Fig. 3.3 Typical phenomena of PW or CW laser welding at low and high power density, showing small molten pool and short plume formation, and tall plume behavior, spattering, deep keyhole, molten pool action, bubble and porosity formation, respectively

and bead welds with CW laser are about 0.5–3 mm deep and 0.5–50 mm deep, respectively. Therefore, shallow laser welds in thin sheets are usually produced by PW laser welding at relatively slow speeds or by CW laser welding at high speeds or under the low power density. It is noted in spot welding with pulsed laser of rectangular wave shape that porosity is likely to be formed in the weld of more than 1 mm depth. Deep welds of thick plates can be produced by using high-power CW laser without wires or with wires. Deep welds are also made by hybrid welding with high-power CW laser and MIG, MAG or CO_2 gas arc.

3.2 Welding with CO_2 Laser

High-power CW CO_2 lasers were first developed to be applied to cutting, drilling, and welding. Photographs of metallic plume and gas plasma observed during CO_2 laser welding of an aluminum alloy A5083 at 5 kW power in He, Ar, and N_2 shielding gas

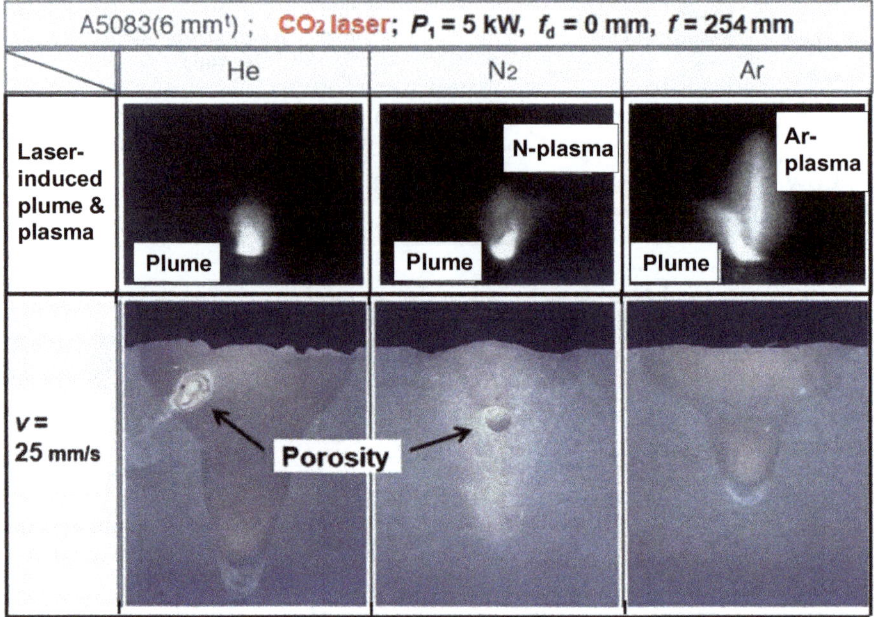

Fig. 3.4 Metallic plume and gas plasma during CO_2 laser welding of Al alloy A5083 at 5 kW in He, Ar and N_2 shielding gas, and consequent geometry or penetration of weld beads at 25 mm/s

and the geometry or penetration of the consequent weld beads are shown in Fig. 3.4 [2]. In He gas, metallic plumes ejected from a keyhole inlet is only observed, and the penetration of a weld bead is the deepest. In N_2 gas, a gas plasma is seen in addition to the metallic plume from the keyhole inlet, and the penetration is slightly shallower than that in He gas at 5 kW power. On the other hand, at the power levels of 2–3 kW deeply penetrated weld beads can be easily formed in N_2 gas probably due to the formation of N plasma and the consequent increase in the laser absorption by forming AlN film covering a molten pool. In Ar gas, a gas plasma is always present under the welding nozzle, and the penetrate is the shallowest. In any shielding gas, porosity is present. As a result of the formation of Ar plasma, it is evident that the Ar plasma has a substantial effect on the reduction in the penetration depth. Nevertheless, it was confirmed at the power of about 3–5 kW that the penetration depth was slightly deeper with an increase in the Ar or He gas flow rate from 15 to 50 L/min for coaxial nozzle of 6–8 mm in diameter. This is understood by considering the effective reduction of a laser-induced plume, a high-temperature zone and especially a laser-induced plasma size in Ar gas due to its high-speed flow during laser welding.

The cross-sectional weld beads obtained in Type 304 austenitic stainless steel at 10 kW in He, He–Ar mixture, Ar and N_2 shielding gas are shown in Fig. 3.5 [4]. The penetration depth is apparently shallower with an increase in the ratio of Ar gas to He gas under the conditions of more than 25% Ar, and the penetration is drastically the shallowest in 100% Ar gas. The weld penetration in N_2 gas is also shallower. Such

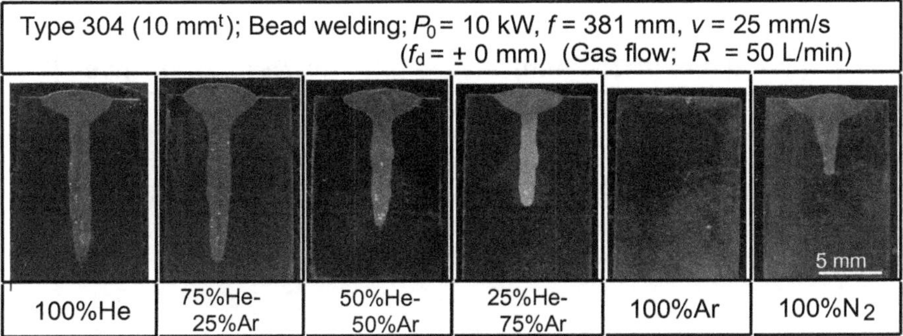

Fig. 3.5 Cross-sectional weld beads made in Type 304 austenitic stainless steel at 10 kW in He, He–Ar mixture, Ar and N_2 shielding gas, showing different weld bead geometries and penetration depths

shallower penetration is attributed to the frequent and periodic formation of Ar or N plasma against the laser beam path. Consequently, He shielding gas is usually used to assure the penetration depth in welding with high power CW CO_2 laser. Besides, N_2 gas is sometimes employed in aluminum alloys at 2–5 kW, because the plasma often facilitates melting of the materials and reduces porosity.

The effect of laser power on the weld penetration in Type 304 steel in He shielding gas was investigated by changing the laser power from 10 to 40 kW. The weld beads obtained is shown in Fig. 3.6 [5]. The penetration depth increased with an increase in the laser power. Porosity was also observed in deep weld beads.

The effect of CO_2 laser beam mode on the weld penetration was investigated by changing the welding speeds. The result is shown in Fig. 3.7 [6]. The effect of beam modes on the weld penetration is apparent, and the penetration is the deepest in a gauss mode, moderate in donut mode and the shallowest in low multi-mode. The difference of the weld penetration is more remarkable at a higher speed.

Such CO_2 lasers were used to weld the parts of pulley, motor cores, gears, etc. High-power CW CO_2 laser welding was first commercially introduced to tailored blank welding, remote welding, welding of aircrafts, welding of thick plates, and so on. These are also first used as a heat source of laser–arc hybrid welding of ships. At present, however, CO_2 lasers are being replaced with disk or fiber lasers in almost all welding applications.

3.3 Welding with YAG Laser

YAG lasers have been used as fiber-delivered PW or CW solid-state lasers. PW and CW lasers were commercially available in the twentieth century, but currently PW lasers only managed to be produced. High-power CW YAG lasers are not commercially available now because these electricity-photon efficiency and beam quality are

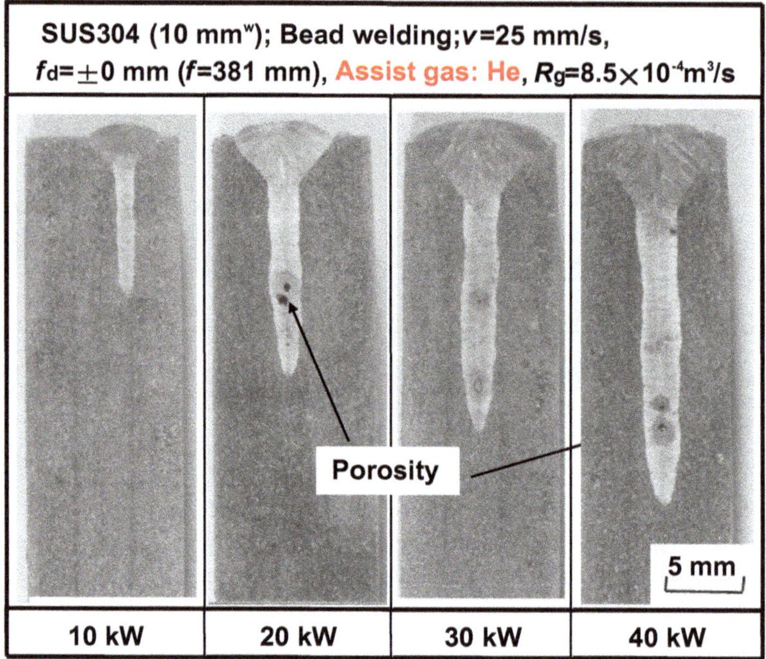

Fig. 3.6 Cross sections of CO_2 laser weld beads made in Type 304 steel at 10–40 kW and 25 mm/s in He gas

much smaller and worse than those of disk or fiber lasers developed mainly in the twenty-first century.

An example of a spot weld made with a pulsed YAG laser is shown in Fig. 3.8, exhibiting the surface, cross section and X-ray inspection result [7]. Type 304 steel plate was welded with the pulsed YAG laser of a rectangular shape. A keyhole type of a spot weld was produced, but bubbles and two pores were formed due to rapid collapse of a keyhole caused by rapid drop in laser power (density).

It is understood from many welding results that spatters and porosity are likely to be formed due to rapid formation of a keyhole in a small molten pool and rapid collapse of a deep keyhole, respectively, in the case of pulsed YAG laser of rectangular pulse shape, as schematically shown in Fig. 3.9 [1]. Pulse-shaped controllable lasers have been available and used to investigate the effect of pulse-shaping on the penetration and welding defects [8, 9]. To suppress spattering and porosity, slow rise and slow fall in the laser power, respectively, are principally needed, as indicated in Fig. 3.9. Especially, a deeper spot weld without porosity could be produced in Al alloy A5083 under the controlled saw-like pulse shape, as indicated in Fig. 3.10 [8]. When the power is raised quickly, spattering occurs due to the rapid formation of a

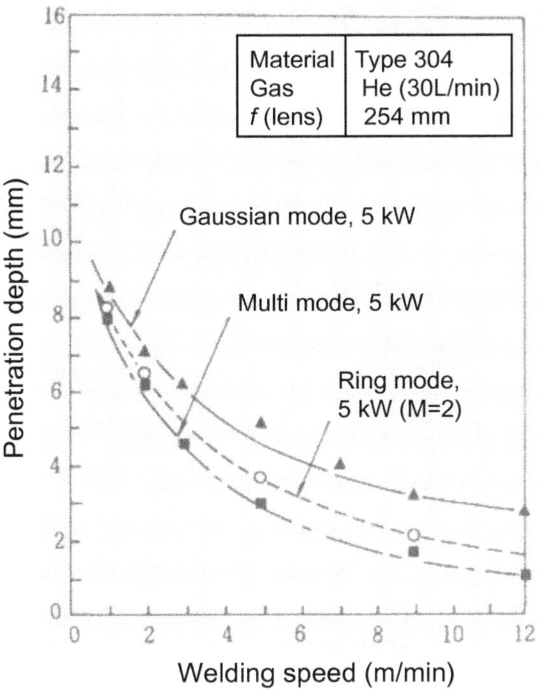

Fig. 3.7 Effect of CO_2 laser beam mode on weld penetration as function of welding speed

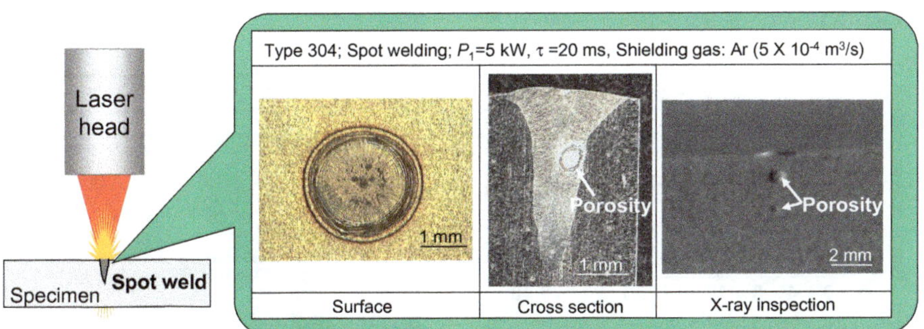

Fig. 3.8 Surface, cross section, and X-ray inspection result of pulsed YAG laser spot weld, showing porosity formation

keyhole in a small molten pool, resulting in the formation of underfilling (concave) surface. A slow rise in the laser power can suppress spattering and a saw-like pulse shape can prevent porosity formation by producing a deep keyhole and the following shallow keyhole one after another.

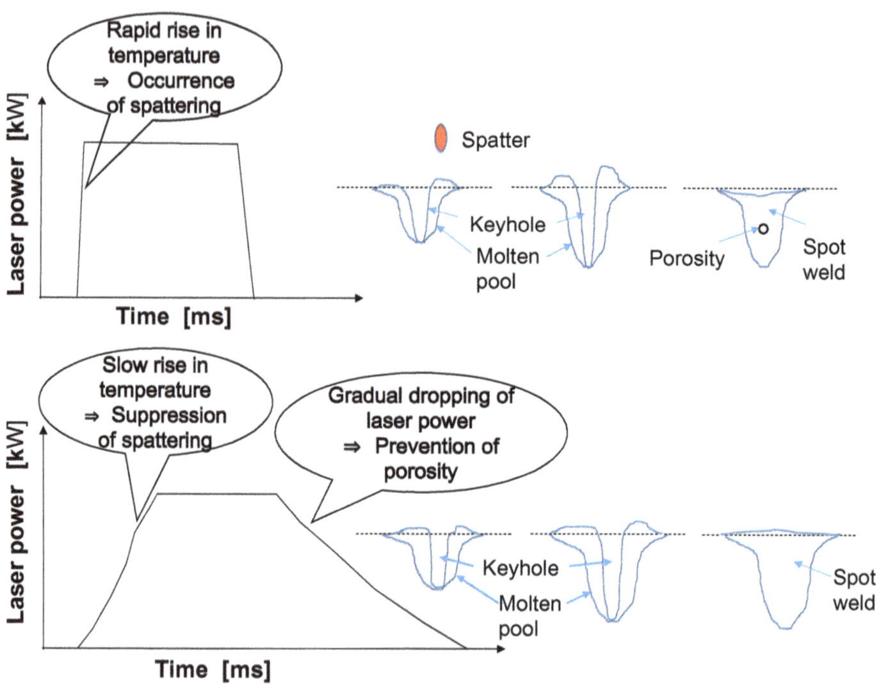

Fig. 3.9 Schematic representation of pulse shapes relating to spattering and porosity formation, showing effect of slow rise and slowdown in laser power on reduction in spattering and porosity, respectively

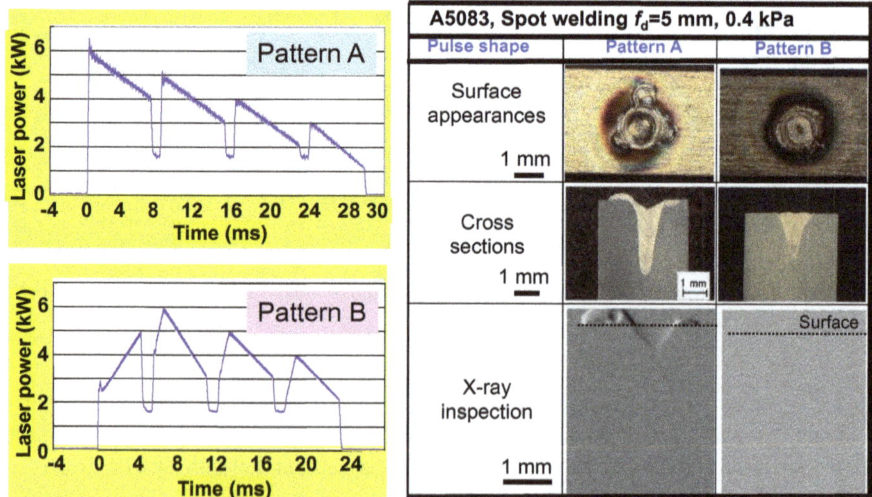

Fig. 3.10 Controlled saw-like pulse shapes of PW YAG laser and deep porosity-free spot welds produced in Al alloy A5083 under their special shapes, showing formation of underfilling weld due to rapid rise in laser power

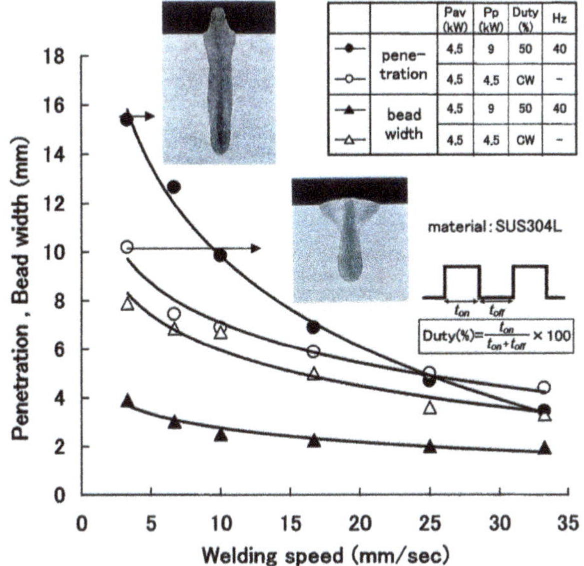

Fig. 3.11 Effects of pulse modulation and welding speed on penetration and width of high power YAG laser weld beads in Type 304 steel

High-power CW YAG laser of 3–6 kW was used to weld Zn-coated steel, stainless steel, aluminum alloys, etc. In the case of the same welding conditions, the weld penetration was deeper when the fiber core diameter was smaller due to higher power density, as shown below in Fig. 3.19 in 3.6. High-power CW YAG lasers of 10 kW were developed, and the effect of pulse modulation to make two times higher power on the weld geometry and penetrations was investigated. The laser welding results of Type 304 steel are shown as a function of welding speed in Fig. 3.11 [10]. The penetration is increased by pulse modulation using higher peak power at low welding speeds. At present, thick plates are subjected to laser welding with a high-power CW fiber or disk laser in place of high-power YAG laser.

Up to now, welding with pulsed YAG lasers has been utilized to produce pacemakers, artificial organs, electronic connectors and components, endoscope parts, power sensor modules, mobile phone batteries, halogen lamps, glasses frames, and so on.

3.4 Welding and Brazing with Disk Laser

High-power CW disk lasers were developed in place of high-power YAG lasers. The penetration depths of disk laser welds in aluminum alloy obtained by using focusing optics of 100 mm and 200 mm in focal length are indicated as a function of welding speed in Fig. 3.12 [11]. It generally indicates that the penetration is deeper at higher welding speed for 100 mm focal length and at lower welding speed for 200 mm focal length.

The effect of welding speed on the geometry of (6 kW) disk laser weld beads made in Type 304 steel in Ar shielding gas is shown in Fig. 3.13 [12, 13]. All laser weld beads are of a keyhole type, and the penetrations are deep. The weld bead penetration decreases with an increase in welding speed. At the welding speed of less than 50 mm/s, porosity is present in deep weld beads. On the other hand, at the welding speeds of 150 mm/s or faster, underfilled weld beads are formed due to severe spattering of large molten droplets. Sound weld beads can be formed at the middle welding speeds of, for example, 75–100 mm/s. Figure 3.14 [13] shows the effect of defocused distance on the geometry and penetration of laser weld beads in Type 304 steel. Spattering is reduced under the distance of −4 mm. Moreover, at the distance of −2 mm the weld bead penetration is the deepest and the underfilled bead is prevented. At +2 mm, however, spattering occurs, and the penetration depth is the shallowest. It is apparent that the defocused distance greatly affects the weld penetration depth and spattering. Spattering can be also decreased with the addition of ring mode, beam forward inclination, smaller laser beam diameter, vacuum conditions, and so on. These results and causes will be described later in detail in the section dealing with welding defects in the Sect. 4.2.

Disk lasers have been used for car bodies, door frames, and parts as well as Li-ion batteries. Disk laser brazing is also applied to car bodies and back sheets.

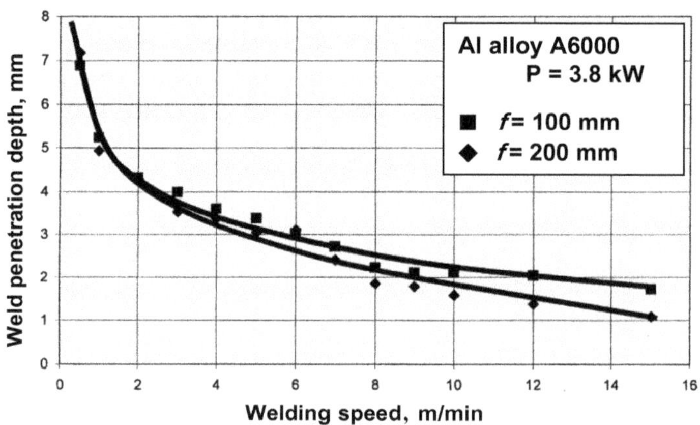

Fig. 3.12 Effects of welding speed and optical focal lengths of 100 and 200 mm on penetration depths of disk laser weld beads made in aluminum alloy

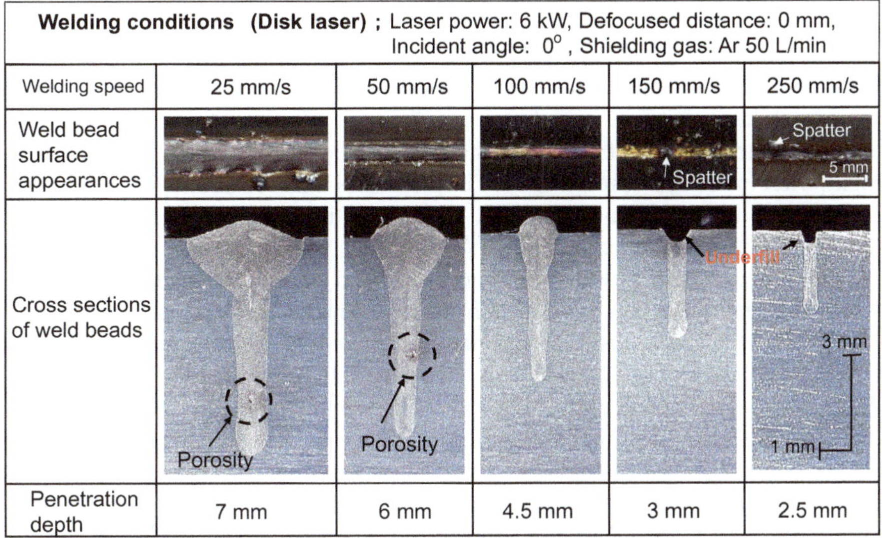

Fig. 3.13 Effect of welding speed on geometry of disk laser weld beads produced in Type 304 steel in Ar shielding gas, showing formation of porosity and underfilling at low and high welding speeds, respectively

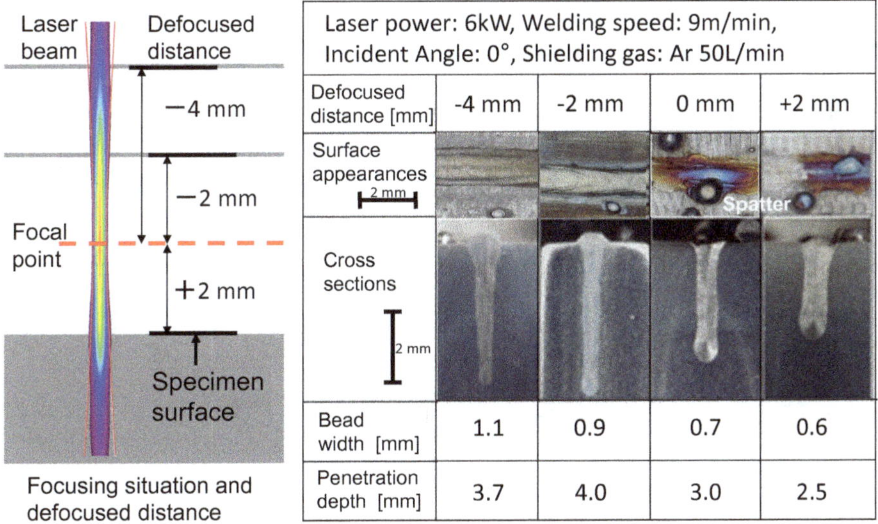

Fig. 3.14 Effect of defocused distance on geometries and penetration depths of laser weld beads in Type 304 steel and formation tendency of spattering during welding

3.5 Welding and Brazing with Fiber Laser

High-power CW fiber lasers were developed almost at the same time or just after high-power disk lasers in place of high-power YAG lasers. It seems that the beam quality and the electric-photon conversion efficiency of fiber lasers are the best among the high-power lasers at the present time. The CW fiber lasers of 120 kW in the maximum power are commercially available, and 10–100 kW fiber lasers are used to investigate the weldability of metallic materials.

The effect of laser power on the geometry and penetration depth of Type 304 steel weld bead at the speed of 3 m/min (50 mm/s) in Ar shielding gas is exhibited in Fig. 3.15 [14]. It is obvious that the penetration increases with an increase in the laser power even in Ar shielding gas although the penetration becomes shallow in welding with CO_2 laser of (10 kW class) high powers in Ar shielding gas. It is understood that the effect of shielding gas or environment is small in high-power fiber laser welding.

The effect of focusing situation on the full-penetration welding results was investigated by welding HT 780 steel plates of 10 mm thickness with the use of two fiber laser optical heads with hot wires at the power of 10 kW and at the speed of 1.5 m/min in Ar shielding gas of 40 L/min. Focusing situations of two focusing optics were measured. Surface appearances, cross sections, and X-ray inspection result of laser weld beads, high-speed video observation of molten pools, and X-ray transmission observation during laser welding were performed. Two optical heads and their focusing situations, and the surface appearances, cross sections, and X-ray inspection results of laser weld beads are compared in Fig. 3.16 [15]. One optic could focus the beam into 200 μm spot diameter at the focal point with the Rayleigh length of 2.5 mm, while the other could obtain the slightly wider spot diameter of about 270 μm and the slightly longer Rayleigh length of 4 mm. Consequently, in the case of smaller spot diameter, the molten pool acted violently and spatters occurred,

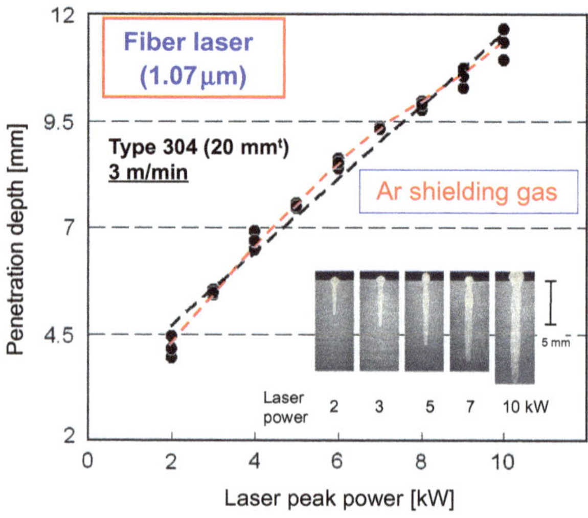

Fig. 3.15 Effect of fiber laser power on penetration depths of weld beads in austenitic stainless steel Type 304 at 3 m/min in Ar shielding gas

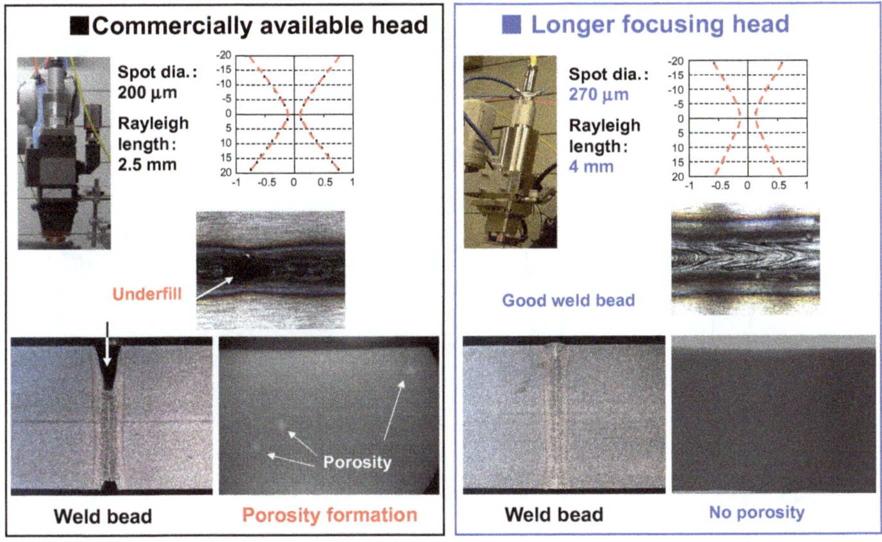

Fig. 3.16 Optical heads, focusing situations, beam spot diameters and Rayleigh lengths, and effect of focusing situation on surface appearances, cross sections, and X-ray inspection results of laser weld beads made at 10 kW, 1.5 m/min, and 40 L/min Ar gas flow rate in HT 780 steel plates of 10 mm thickness

resulting in the formation of an underfilled bead with porosity. On the other hand, in the case of the slightly wider spot diameter, the molten pool behaved calmly and no bubbles were generated, resulting in the formation of a sound full-penetration weld bead without underfill nor porosity. It is concluded that the proper selection of a focusing optic and situation is extremely important in producing sound laser weld beads.

The penetrations and defects of weld beads were investigated in welding with 100 kW fiber laser. An example of a fully penetrated weld bead produced in butt joint of Type 304 steel of 70 mm thickness by two passes at 100 kW and 2 m/min is shown in Fig. 3.17 [16]. A weld bead of about 40 mm penetration depth could be produced at 100 kW and 2 m/min in Ar shielding gas. And thus, 70-mm-thick plates could be satisfactorily welded by two passes [17]. In low vacuum of 5 kPa, a sound full-penetration weld bead of two passes could be produced in Type 304 steel plate of 150 mm thickness at 60 kW and 0.3 m/min, as shown in Fig. 3.18 [17]. Sound one-pass full-penetration weld beads could also be formed in 40-mm-thick steel plates at 50 kW at the slow speed of 0.3 m/s in air atmosphere at 1 atm by using flux and ceramic plate on the top and bottom surfaces, respectively [18].

Fiber lasers are also employed to brazing of Zn-coated steel sheets. Especially, fiber lasers of Mickey Mouse type of three spot beams are developed and used to stably produce a sound brazed joint by forming small molten pools with heat sources corresponding to two ears diagonally in right and left front of the main molten pool [19].

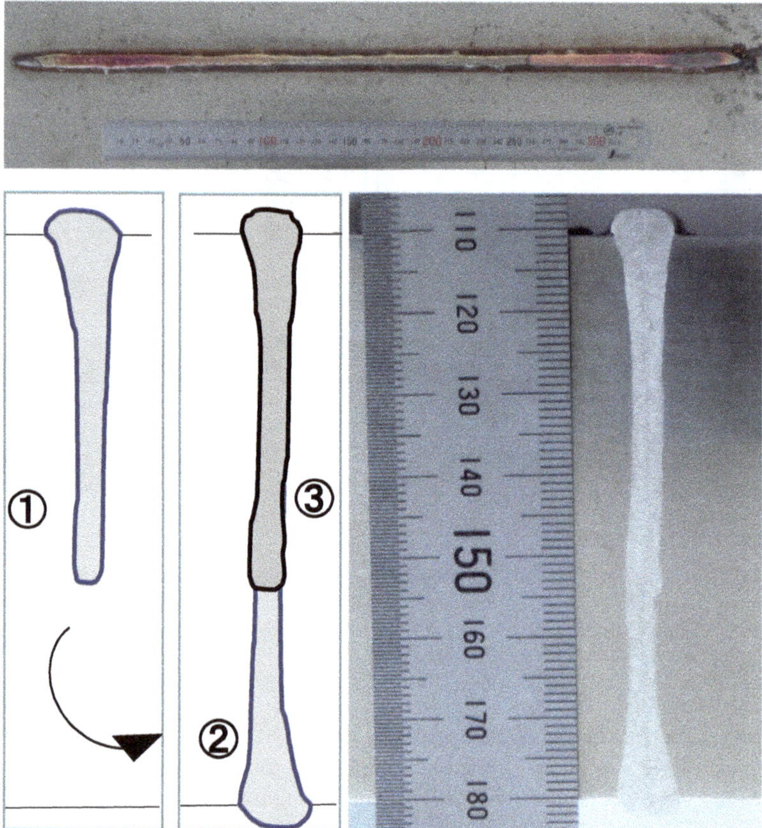

Fig. 3.17 Example of fully penetrated weld bead in butt joint of Type 304 steel plate of 70 mm thickness made with 100 kW fiber laser by two passes at 2 m/min in Ar shielding gas

Fiber lasers emitting in a pulse shape are also commercially available. These are expected to be utilized in the place of pulsed YAG laser.

Fiber lasers are used in almost all industrial fields and are mostly expected as industrially applicable lasers.

3.6 Welding, Brazing, and Soldering with Diode Laser

High-power diode lasers of less than 60 kW have been developed and are commercially available now. Such high-power diode lasers of less than 4 kW were first used by directly holding the devices on the robot arm or by direct irradiation due to the setting of a device and focusing optic but presently are chiefly utilized by fiber delivery of the laser beam. In the latter case, the fiber cores are normally of

(a) Bead surface

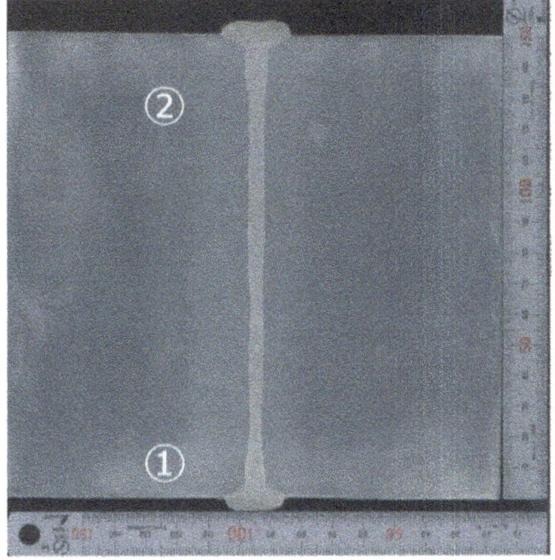

(b) Cross section

Fig. 3.18 Example of sound full-penetration fiber laser weld bead of two passes produced in Type 304 steel plate of 150 mm thickness at 60 kW and 0.3 m/min in low vacuum (5 kPa)

0.6–1 mm large diameter although such diameters correspond to the conventional high-power YAG lasers. The other type of diode lasers with the fiber cores of 0.2 mm small diameter have been developed, and now high-power lasers of less than 8 kW are commercially available.

The geometries and penetration depths of weld beads of direct diode laser of a rectangular shape were compared with those of YAG lasers of 0.6 and 1 mm in focused beam diameter. The results are shown together with the observation results of plume and spattering during welding in Fig. 3.19 [20]. The penetration depths of diode laser welds were shallower than those of YAG laser welds because of lower power density. Nevertheless, a keyhole type of a diode laser weld bead was obtained at the power of 4 kW and the welding speed of 1 m/min. Besides, it was observed that spattering could be drastically reduced with a rectangular mode of diode laser in

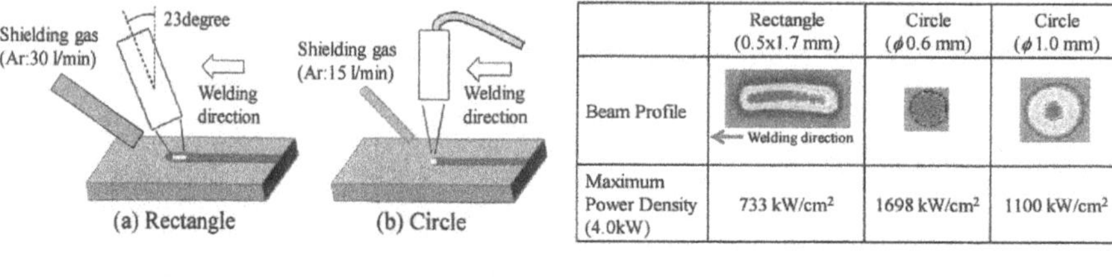

	Rectangle (0.5x1.7 mm)	Circle (ϕ0.6 mm)	Circle (ϕ1.0 mm)
Beam Profile			
Maximum Power Density (4.0kW)	733 kW/cm²	1698 kW/cm²	1100 kW/cm²

	Welding Speed(Laser Power:4.0 kW)		
	1.0m/min	4.0m/min	6.0m/min
Rectangle (0.5x1.7mm)	2mm	2mm	2mm
Circle (ϕ0.6mm)	2mm	2mm	2mm
Circle (ϕ1.0mm)	2mm	2mm	2mm

	Welding Speed(Laser Power:4.0kW)		
	1.0m/min	4.0m/min	6.0m/min
Rectangler (0.5x1.7mm)			
Circular (ϕ0.6mm)			
Circular (ϕ1.0mm)			

⟵ Welding Direction

Fig. 3.19 Comparison of geometries and penetration depths of weld beads made with direct diode laser of rectangular shape and YAG laser of 0.6 mm diameter and 1 mm diameter

comparison with YAG lasers. It is beneficial to know the great effect of a rectangular mode laser on the reduction in spattering.

High-power diode lasers of 50 and 60 kW are commercially available and are used to produce deeply penetrated weld beads, as the 25-mm-thick or 20-mm-thick steel plates are subjected to butt-joint welding and T-joint welding with a 50 kW diode laser at 1 m/min, as shown in Fig. 3.20 [21]. A hot crack is apt to occur in a

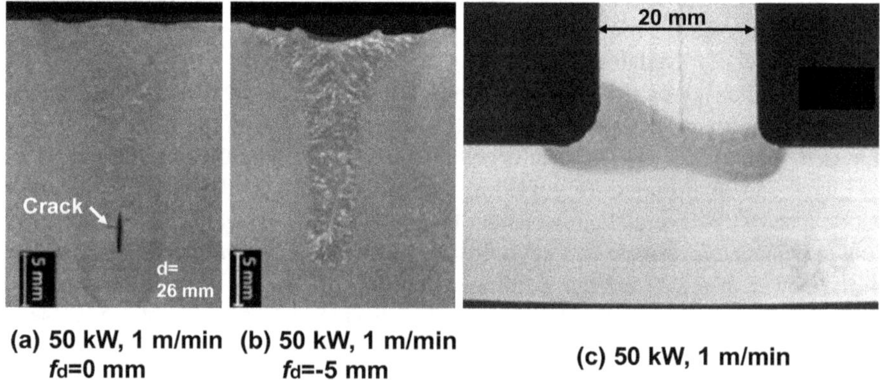

Fig. 3.20 Cross sections of butt-joint weld beads made with 50 kW diode laser at 1 m/min and defocused distance of 0 mm (**a**) and -5 mm (**b**), showing crack and no crack, respectively, and example of sound T-joint weld bead produced with high-power diode laser (**c**)

partially penetrated deep weld bead. Fairly good weld beads can be produced with a high-power diode laser under the proper conditions.

High-power diode lasers are used as heat sources for fillet welding of an aluminum alloy sheets in car doors and for brazing of Zn-coated steel sheets in car bodies or trunk lids, as shown in Fig. 3.21 [22]. Good weld beads and beautiful brazing joints can be produced with diode lasers at high welding speeds.

Diode lasers of low powers are utilized as heat sources for soldering. Compact and low-cost machines of diode lasers are available. Connector and lead wires, ICs, display information processor (DIP) resistors are manufactured by diode laser soldering.

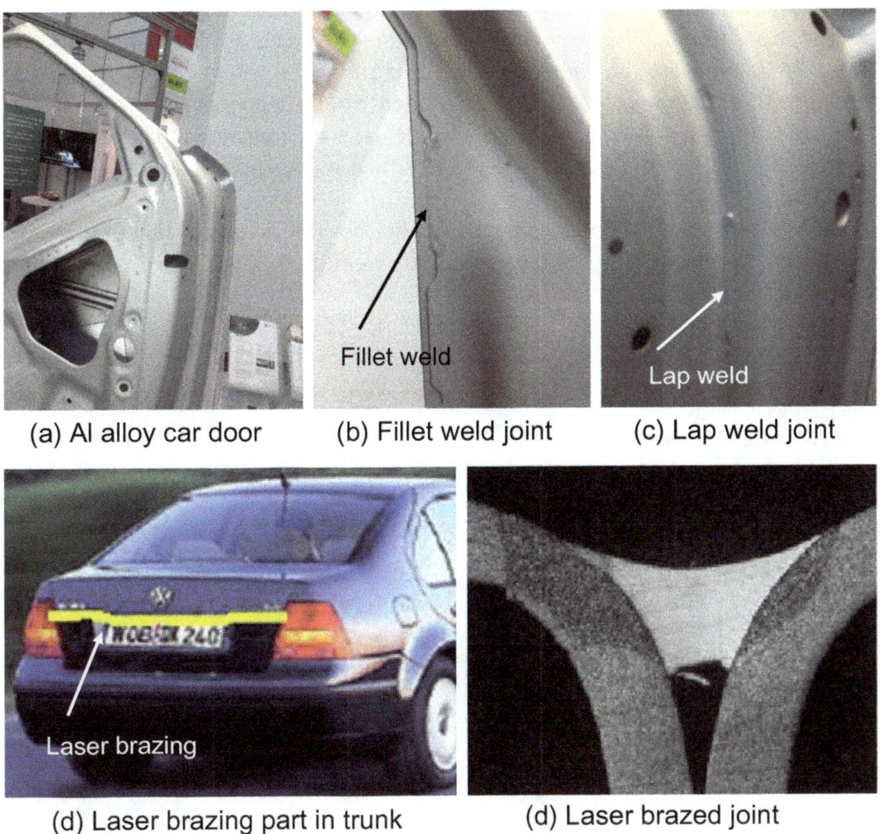

Fig. 3.21 Examples of fillet and lap welds of aluminum alloy sheets in car doors and brazing joint of Zn-coated steel sheets in car trunk lid

Diode lasers of wide power ranges of 20 W to about 4 kW are used for joining of similar or dissimilar plastics or plastic to metal. Actually, some parts of cars, etc., are manufactured by joining plastics with diode lasers.

3.7 Welding and Soldering with Green or Blue Laser

High-power CW green and blue lasers are obtained as second harmonic generation (SHG) lasers of disk lasers and diode lasers, respectively. Pulsed green lasers are also available from YAG, disk and fiber lasers. The absorptivity of green or blue lasers into Cu (copper) plate is higher than that of other lasers, and thus, these lasers are used to understand the advantages of welding and joining of Cu sheets. The cross sections of blue laser weld beads melted in Cu sheets at 275 W & 5 mm/s and 500 W & 10 mm/s are shown in Fig. 3.22 [23]. The sound weld beads suggesting the formation of a shallow concavity can be formed. Sound deeper penetration welds are obtained with high power green lasers. Besides, it is demonstrated that spot welds of green lasers of disk or YAG lasers can be formed more stably than those of pulsed YAG lasers [24].

High-power green and blue lasers are applied for stable lap welding of many thin Cu sheets, hairpin welding of Cu, dissimilar welding of Cu to steel or aluminum alloy sheets, and so on.

Fig. 3.22 Cross sections of blue laser weld beads produced in Cu sheets at 275 W and 5 mm/s and 500 W and 10 mm/s

At present, the powers of blue lasers are less than 2 kW and are not high, and thus, hybrid systems of 1–1.5 kW blue laser of 1 mm core diameter and 1 mm spot diameter at the focal point and 1–7 kW IR laser of 0.6 mm core diameter and 0.2–0.3 mm spot diameter are developed to apply laser welding of Cu sheets [25]. In the case of a blue laser only, the weld penetration is shallow because of low power and low power density. In the case of an IR laser, an unstable weld bead is formed, while stable full-penetration weld beads are produced in hybrid welding. It is concluded that the stable weld beads can be produced in Cu plate by making a good use of the hybrid heat source of blue laser and IR laser.

Pulsed blue lasers of low powers are also used as heat sources for laser soldering. More stable soldering for the formation of sound joints is achieved with blue diodes than with normal diode lasers of about 0.8–1 μm in wavelength.

3.8 Welding with Picosecond Laser or Femtosecond Laser

Femtosecond (fs) lasers are characterized by extremely short pulse durations and super high peak power. Therefore, these are applied to precision processing of semiconductors and liquid crystals, processing of transparent materials such as glasses and sapphires, processing of engine parts in cars or aircrafts, processing of optical communication components, surface hardening of metallic materials, and so on.

Joining of glasses has been performed with CO_2 laser. Recently, lap joining of glasses is tried by using femtosecond lasers. It is revealed that lap joining of glasses with picosecond or femtosecond lasers is feasible, as an example of borosilicate glass joining with fs laser (360 fs, 1.045 μm, 1 MHz) is shown in Fig. 3.23 [26,

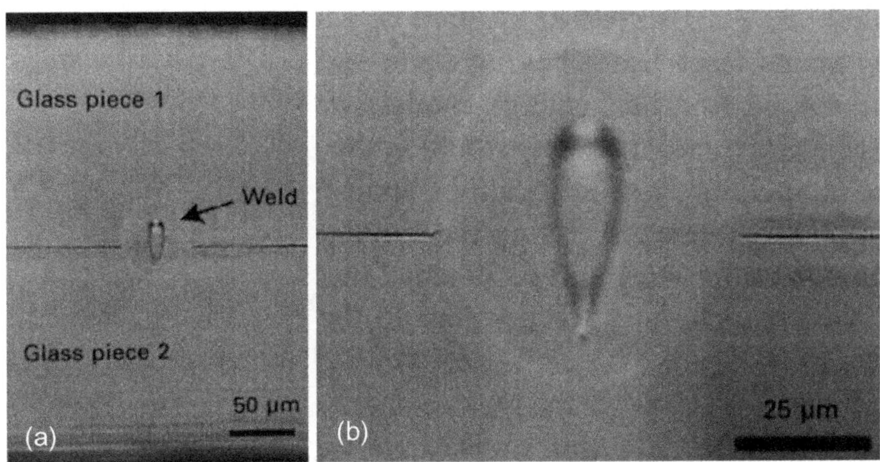

Fig. 3.23 Example of lap joining of borosilicate glasses produced with femtosecond laser (360 fs, 1.045 μm, 1 MHz). **a** Cross section of lap joint, **b** magnified photograph of lap joint

27]. Damaged areas may be formed when the power density is too high, though. Multipath overlap welding of borosilicate glass with a picosecond laser of 10 ps was carried out with a separation of 15 μm, and the result demonstrates that a strong joint could be produced.

References

1. Katayama S (2019) Very easy book of laser processing. The Nikkan Kogyo Shimbun Ltd. (in Japanese)
2. Katayama S (2013) Introduction: fundamentals of laser welding. In: Katayama S (ed) Handbook of laser welding technologies. Woodhead Publishing Limited, pp 3–16
3. Katayama S (2012) Ferrum (Bulletin of the Iron and Steel Institute of Japan) 17(1):18–29 (in Japanese)
4. Seto N, Katayama S, Matsunawa A (2000) J Laser Appl 12(6):245–250
5. Katayama S, Seto N, Kim JD, Matsunawa A (1997) In: Proceedings of ICALEO '97, LIA, vol 83-Part 2, Section G, pp 83–92
6. Personal Communication (2000) Mitsubishi Electric Corp.
7. Katayama S, Seto N, Mizutani M, Matsunawa A (2001) In: Proceedings of ICALEO 2001, LIA, Session C: Welding, 804 (CD), p 804
8. Katayama S, Mizutani M, Ikeda H, Nishizawa K, Matsunawa A (1992) In: Proceedings of ICALEO '92, LIA, pp 547–556
9. Katayama S, Kohsaka S, Mizutani M, Nishizawa K, Matsunawa A (1993) In: Proceedings of ICALEO '93, LIA, pp 487–497
10. Shimokusu Y, Fukumoto S, Nayama M, Ishide T, Tsubota S, Matsunawa A, Katayama S (2002) J Jpn Weld Soc (JWS) 20(4):477–483 (in Japanese)
11. TRUMPF + IFSW, University of Stuttgart Data
12. Kawahito Y, Katayama S et al (2015) J Jpn Weld Soc (JWS) 33(1):13–19 (in Japanese)
13. Kawahito Y, Nakada K, Uemura Y, Mizutani M, Nishimoto K, Kawakami H, Katayama S (2016) J Jpn Weld Soc (JWS) 34(4):239–248 (in Japanese)
14. Kawahito Y, Mizutani M, Katayama S (2009) Sci Technol Weld Join 14(4):588–594
15. Katayama S (2008) Report of the AMADA foundation, pp 240–245. http://www.amada-f.or.jp/r_report2/kkr/24/AF-2008218.pdf (in Japanese)
16. Katayama S, Mizutani M, Kawahito Y, Makino Y, Kono W, Ito S, Sumimori D (2014) IIW Com. IV, Seoul, Doc. IV-1182-14 (Web)
17. Katayama S, Mizutani M, Kawahito Y, Ito S, Sumimori D (2015) Lasers in Manufacturing, LiM 2015, Munich, Germany, USB (8 p)
18. Nomura R, Sumimori D, Ashida Y, Deguchi T, Watanabe Y, Katayama S (2017, Spring) Reprint Jpn Weld Soc (JWS), Session ID: 201
19. Darvish M, Esen C, Gurevich E, Mamerow H, Ostendorf A (2017) LiM 2017 Joining (Welding & Brazing) (Website)
20. Tokunaga H, Yasuyama M (2012) In: Proceedings of the 78th Japan Laser Processing Society, JLPS, pp 41–45 (in Japanese)
21. Reisgen U, Olschok S, Weinbach M, Engels O (2017) LiM 2017 Joining (Welding & Brazing) (Website)
22. Beyer E (2018) (Fraunhofer IWS), Personal Communication, Laser Market Place 2003
23. Nuburu (The Blue Laser Company) (2018). https://nuburu.net
24. Nakamura T (2018) The latest laser welding system of TRUMPF. In: Proceedings of the 89th Laser Materials Processing Conference, Osaka, Japan, JLPS, vol 89, pp 55–60 (in Japanese)

25. S. Takeda: Personal communication, Laserline (2019). https://www.laserline.com/
26. Miyamoto I (2013) Laser welding of glass. In: Katayama S (ed) Handbook of laser welding technologies. Woodhead Publishing, pp 301–331
27. Bovastek J, Arai A, Schaffer CB (2005) In: Proceedings CLEO/Europe—EQEC 2005

Chapter 4
Laser Welding Results and Phenomena

4.1 Spot Welding with Pulsed Laser

4.1.1 Phenomena during Laser Spot Welding

Spot welding with a pulsed (PW) laser is performed under various conditions to produce a heat conduction type of a shallow weld or a keyhole type of a deep weld depending on the power density according to the defocused distance, pulse duration, pulse energy, and peak power. A schematic of spot welding phenomena is already represented in the left side of Fig. 3.3 in Sect. 3.1. When the laser power density is low, a heat conduction type of a shallow weld is formed. On the other hand, a keyhole type of a deep weld is produced at a high-power density or high-energy density. The behavior of a keyhole and a molten pool during pulsed YAG laser spot welding has been observed by a high-speed video camera. An example of the observation results during spot welding of aluminum alloy A5083 (containing 4.6 mass% Mg) exposed to PW YAG laser of 10 ms pulse duration in air is exhibited in Fig. 4.1 [1, 2]. A small keyhole is observed at 2.07 ms after the initiation of laser irradiation, exists in various irregular shapes during laser irradiation up to 10 ms, and is also present at 10.62 ms after the termination of laser shooting probably due to the surface tension of liquid. An irregular shape of a keyhole is formed in A5083 molten pool covered with oxide films. Such keyholes are stably formed in the round shape during pulsed YAG spot welding of Ti and stainless steel.

A keyhole is formed due to the evaporation and its recoil pressure, and consequently, a keyhole is occupied with evaporated metallic atoms, or partly ions or clusters. These substances are ejected from the keyhole inlet and are seen as a laser-induced plume, or a bright light source. This metallic plume of bright light was conventionally called a laser-induced plasma. Nowadays, the term "plume" is recommended instead of "plasma" since there are few ions nor electrons in spot welding with a pulsed laser. Laser-induced plume behavior is observed by high-speed video

S. Katayama, *Fundamentals and Details of Laser Welding*,
Topics in Mining, Metallurgy and Materials Engineering,
https://doi.org/10.1007/978-981-15-7933-2_4

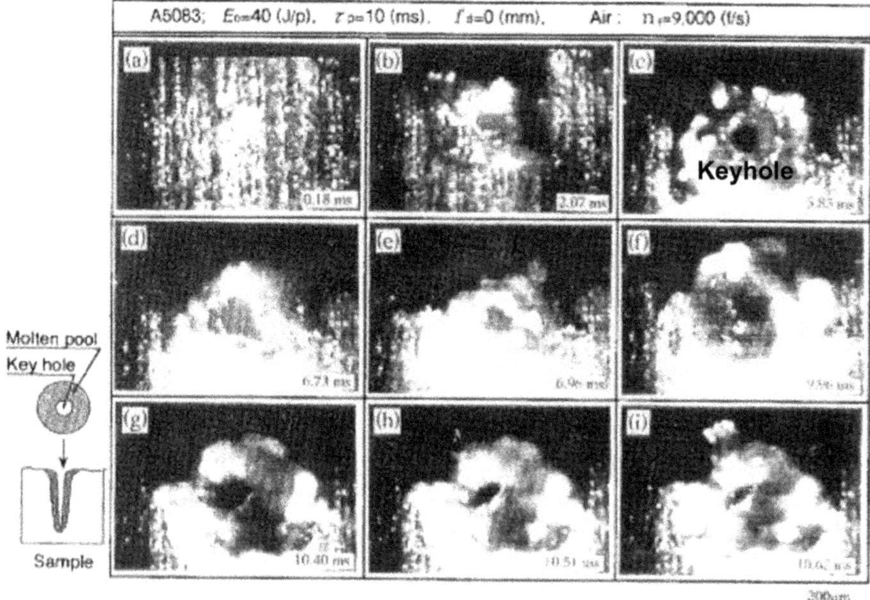

Fig. 4.1 High-speed video observation pictures during spot welding of aluminum alloy A5083 containing 4.6 mass% Mg exposed to YAG laser of 10 ms pulse duration in air, showing keyhole formation in molten pool

camera and high-speed streak camera. An example of observation results is given in Fig. 4.2, showing high-speed streak image and frame photographs during spot welding of pure Ti with pulsed YAG laser of 28.7 J/p pulse energy and 7 ms pulse duration in air environment [1]. It is shown that the plume first grows up stably from the irradiated molten pool and the initial cavity, and then, it evolves from the keyhole inlet repeatedly in the periodic motion. Under these conditions, the propagation speed of a plume is about 20 m/s, and the periodic motion is 210–320 μs and approximately 230 μs in average (for 2.5–4 ms).

An emission spectrum of a laser-induced plume was measured by multi-channel spectrophotometry system to know the effect of laser power density on evaporation behavior and its emission intensity. Examples of the measurement results during pulsed YAG laser spot welding of aluminum alloy A5083 in air are shown in Fig. 4.3 [1, 3]. When the power density is low, Mg and MgO emission is predominant. This is attributed to the high vapor pressure of Mg at low temperatures. Mg becomes MgO after oxidation reaction with O in air above the keyhole inlet. As the power density is increased, the emission from AlO appears and becomes stronger. When the power density is higher, Al of a higher content is largely evaporated, and Al is oxidized to AlO reacted with oxygen (O) in air. In the case of the metallic materials in Ar atmosphere, metallic atoms were detected by preventing oxidation. Ions emission lines were not detected during pulsed YAG laser spot welding of various materials,

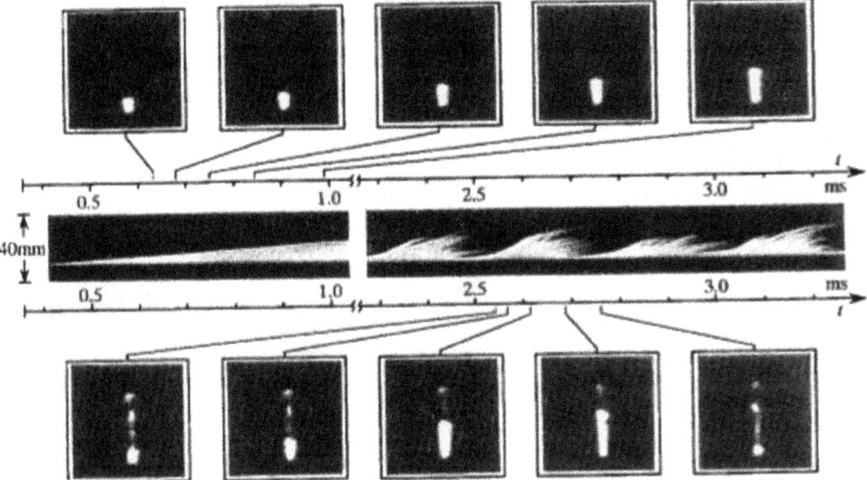

Fig. 4.2 Photographs of laser-induced plume behavior above keyhole inlet of Ti molten pool observed by high-speed video camera and high-speed streak camera, showing periodic growth and evolution of plume in initial stage of laser irradiation

which supports the technical term "a laser-induced plume" instead of "a laser-induced plasma."

Ultrafine particles composing of atoms evaporated during pulsed YAG laser spot welding are collected and observed by transmission electron microscope (TEM) and analyzed by EDX or EDS (energy-dispersive X-ray spectroscopy). The result of ultrafine particles from A5083 alloy is shown in Fig. 4.4, presenting TEM photograph, electron diffraction pattern and its key diagram, and EDS spectra. Ultrafine particles of $MgAl_2O_4$ oxides are formed during spot welding in air, and moreover, the ultrafine particles of mainly Al and partly Mg are produced in Ar atmosphere [1, 4]. That is to say, metallic ultrafine particles are formed in Ar gas although oxide ultrafine particles are produced in air.

The effect of a high-temperature zone composing of a laser-induced plume and ultrafine particles on the incident laser beam was studied by investigating the interaction against the other probe laser. The measurement setups and results during pulsed YAG laser spot welding of Ti (titanium) are shown in Fig. 4.5 [5]. The decreased transmittance of probe laser power is measured, and simultaneously scattered laser is detected. Consequently, the cause of the attenuation of laser power is attributed to Rayleigh scattering due to ultrafine particles.

The above observation and analytical results are conducive to better understanding of the phenomena during spot welding with a pulsed laser, as shown already in Fig. 3.3. A keyhole grows downward by the recoil pressure due to the plume ejection upwards, and the plume is composed of exited atoms emitting bright light and forms a high-temperature zone or region. Such a long zone of high-temperature vapors acts

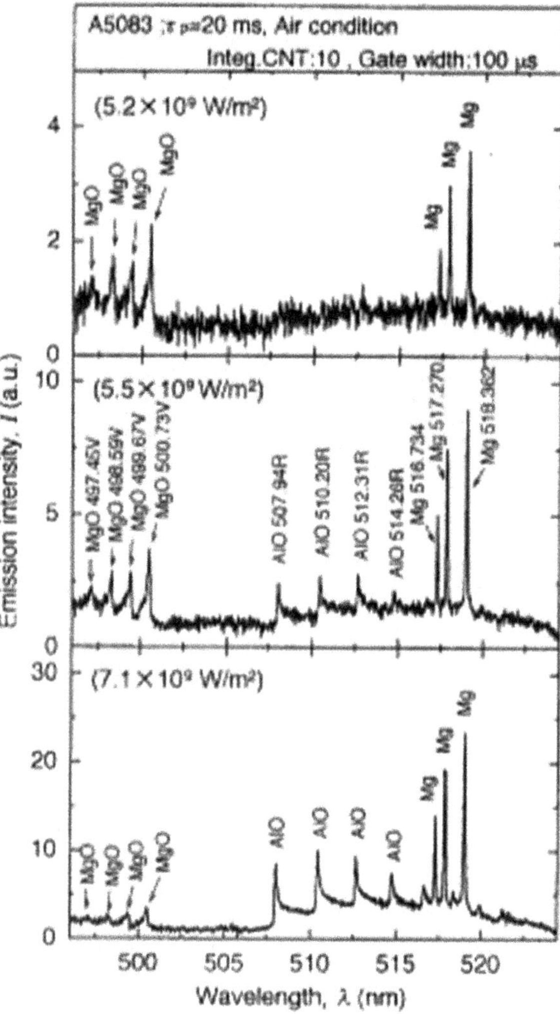

Fig. 4.3 Analytical multi-channel spectrophotometry results of emission spectrum of laser-induced plume from A5083 plate exposed to YAG laser for 20 ms pulse duration at different power densities

to shift the focal position downwards and refracts the laser beam. Afterward, the vaporized atoms gather to form clusters or ultrafine particles, leading to fumes.

Clusters and ultrafine particles existing within the laser incident path act as sources of Rayleigh scattering.

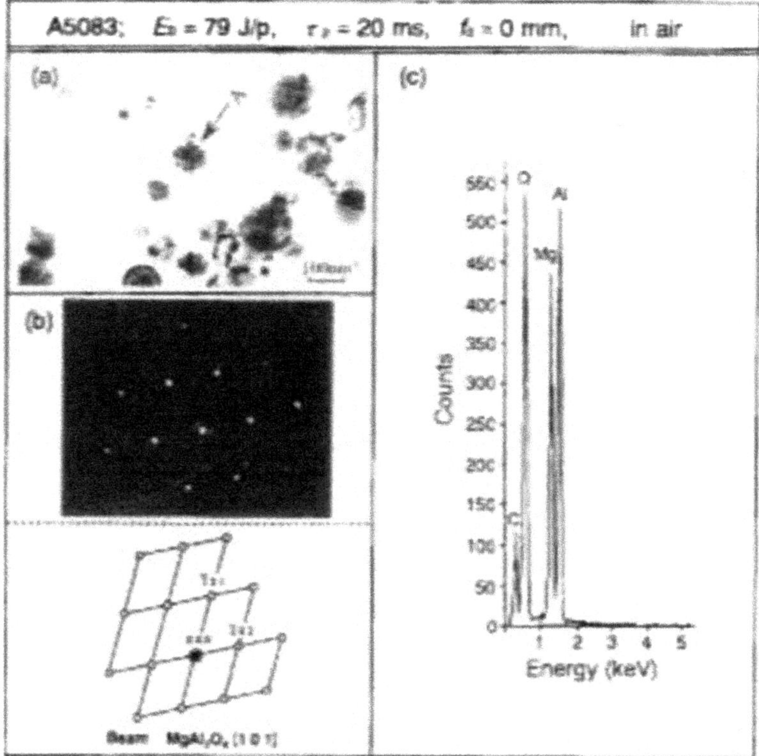

Fig. 4.4 TEM photograph, electron diffraction pattern and its key diagram, and EDX result of ultrafine particles formed in pulsed YAG laser welding of A5083 alloy, showing formation of $MgAl_2O_4$ oxide ultrafine particles (The film holding ultrafine particles is made of C (Carbon), and consequently C is detected in EDX result in **c**)

4.1.2 Effect of Laser Welding Conditions on Weld Penetration and Defects

Spot welding is performed to produce small welds with a pulsed laser in different shielding gases under various conditions of different defocused distances, different laser pulse energies with different pulse durations and different powers. Spot welding was carried out with pulsed YAG laser of roughly rectangular shapes at the different defocused distances for respective pulse durations of 1, 2, 3, 5, and 10 ms. Schematic cross-sectional spot welds with or without porosity obtained are summarized in Fig. 4.6 [6]. The respective pulse shapes used are also indicated. The weld penetration increases with approaching the focal point at respective pulse durations and with increasing the pulse duration at the same defocused distance. Porosity is present in the case of shorter defocused distances, shorter pulse duration, and deeper welds.

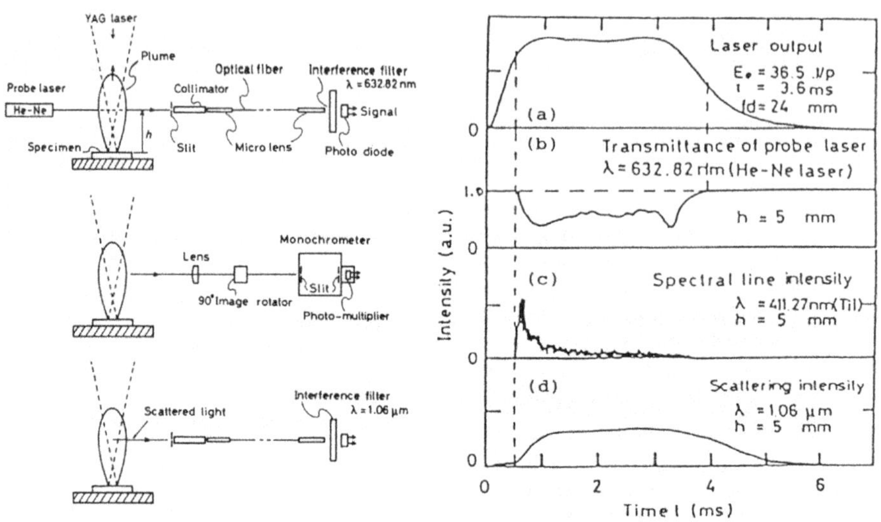

Fig. 4.5 Schematic setup for measurement of probe laser attenuation, variation of spectral line intensity of Ti(I) and scattering intensity, and their measurement results during pulsed YAG laser spot welding of Ti (titanium)

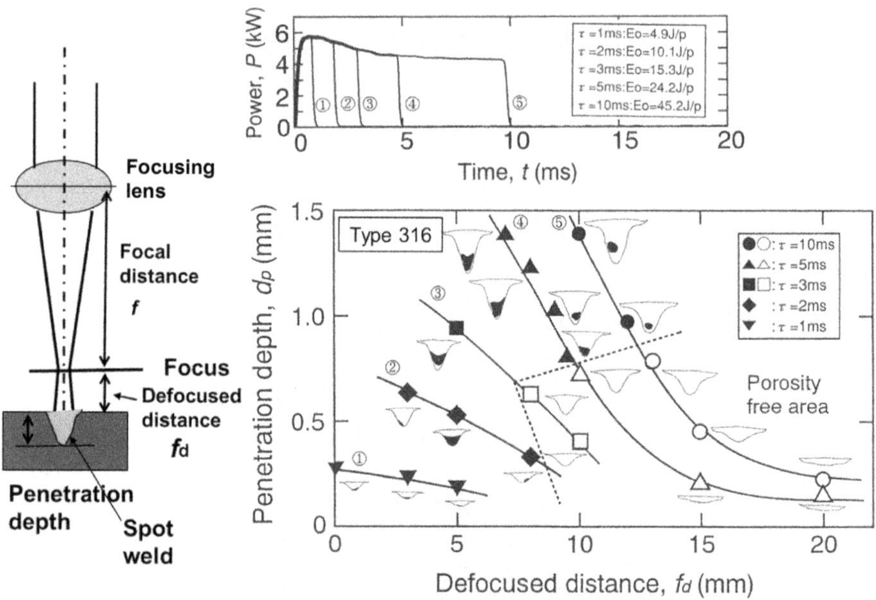

Fig. 4.6 Schematic illustration of defocused distance, various rectangular pulse shapes of 1–10 ms in pulse duration used, and schematic cross-sectional spot welds in Type 316 steel, showing penetration depths of laser spot welds as function of pulse duration under defocused conditions, and effects of defocused distance, pulse duration and penetration depth on porosity formation

This suggests that sound good spot welds can be obtained at the depth of less than 0.8 mm in the case of the rectangular pulse shape with the width of more than 3 ms.

Keyhole behavior during spot welding with a pulsed YAG laser was observed through microfocused X-ray transmission imaging system. The observation results of keyhole behavior during spot welding with a laser of 20 ms pulse duration are shown in Fig. 4.7 [7]. A keyhole was formed about 3 ms after initiation of laser shooting due to the time required for heating and melting and was present about 1 ms after termination of laser irradiation probably due to the surface tension of a keyhole wall. The porosity formation was observed after laser termination. From such observation results, it is interpreted that bubbles leading to porosity are formed from part of a keyhole, as schematically illustrated in Fig. 4.8 [7]. Porosity formation

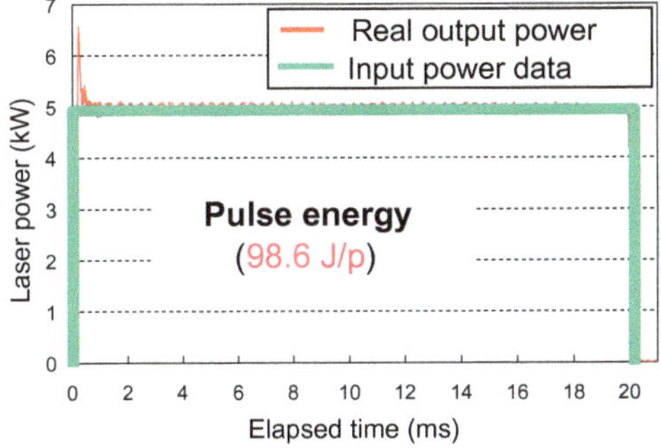

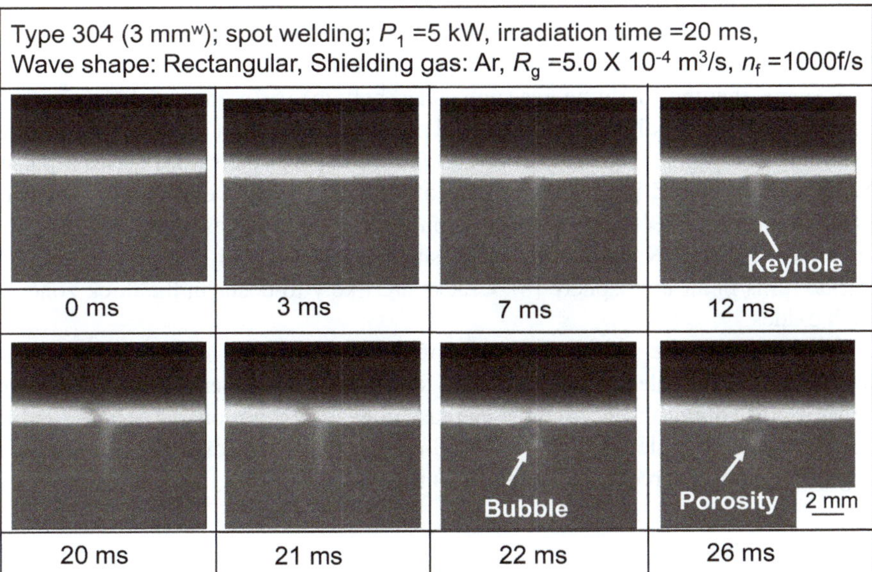

Fig. 4.7 Pulse shape used and pre-setup pulse shape, and pictures showing keyhole behavior and bubble generation leading to porosity formation during spot welding with pulsed YAG laser observed through microfocused X-ray transmission in situ observation system

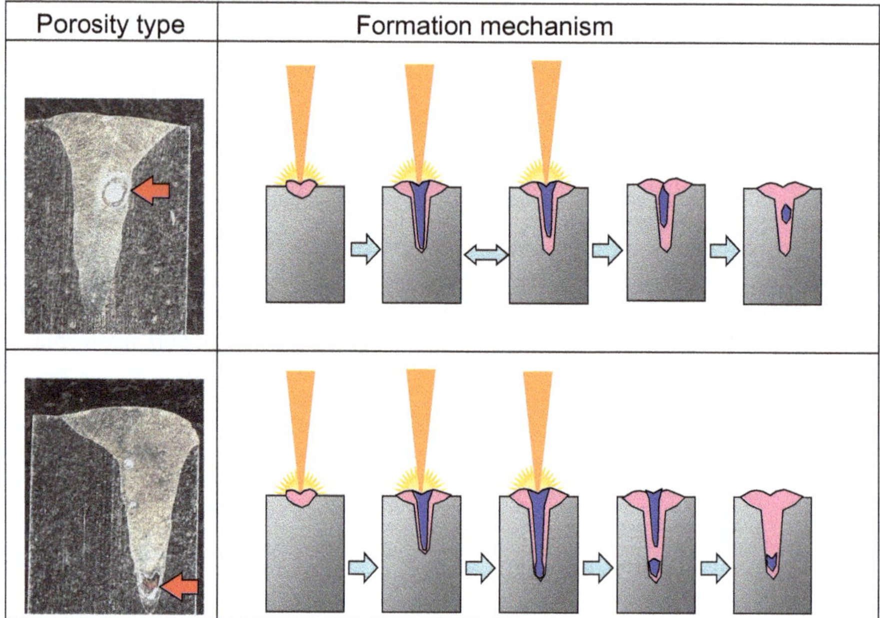

Fig. 4.8 Porosity (pores) in spot welds, and schematic formation mechanisms of keyhole and bubbles, leading to porosity formation near middle and bottom of spot welds

is attributed to the rapid reduction in pulsed laser power and the formation of a deep keyhole. A shielding gas is included in such porosity.

It seems that the shielding gas such as He(helium), Ar(argon), and N_2(nitrogen) hardly affect the welding phenomena and penetration depth in spot welding of most materials except for Ti (titanium) or Zr (zirconium) and their alloys with a pulsed YAG laser. In pulsed laser spot welding of Ti or Zr, sound welds can be produced in He or Ar gas under the proper conditions, while the surfaces of the weld nuggets are covered with TiN nitride film in N_2 shielding gas or with TiO_2 oxide and TiN nitride films in air atmosphere [8]. Cracks are likely to occur in the thick zones of such oxides.

In conclusion, the penetration depth of a laser spot weld mainly depends upon the depth of a keyhole formed according to the material properties, the laser power density, pulse duration, etc., and the increase in weld penetration due to heat conduction from the keyhole tip after the termination of laser irradiation is limited to a negligible small degree, for example, less than 0.1 mm.

4.1.3 Effect of Laser Pulse Shaping on Weld Penetration and Defects

Sound spot welds can be produced in metallic materials, but spattering resulting sometimes in a concave surface occurs during spot welding, porosity is easily formed in deep spot welds, and hot cracking or especially solidification cracking takes place in most alloys. Such welding defects should be reduced or prevented by selecting the proper welding conditions and by specially controlling the pulse shapes.

The main spatters are formed in the initial stage of laser irradiation when the power is raised rapidly, and subsequently, several small spatters are ejected during laser irradiation. When the main pulse laser with pre-pulse small power is irradiated, severe spattering is suppressed because a molten pool is formed by pre-pulse laser and a keyhole is generated in the molten pool. Then, spattering can be suppressed by accommodating the expansion due to the growth of a keyhole in the molten pool. Such results of Ti welds are compared in Fig. 4.9 [9]. Large spatters are present on the plate surface in the case of rectangular pulse shape. In fact, it is observed by high speed cameras that spatters can be reduced. On the other hand, porosity as one of the welding defects is present in both spot welds because of the formation of a deep keyhole and probably rapid collapse of a keyhole due to rapid falling of laser power The larger overlapping ratio of spot welds can reduce porosity, and moreover, a slow decrease in the laser power can decrease porosity appreciably, as shown in Fig. 4.10 [9]. X-ray observation during spot welding exhibited no bubble formation resulting in no porosity.

The effect of laser pulse shaping or power decrease in pulsed YAG laser spot welding of stainless steel is shown in Fig. 4.11 [1, 6]. The porosity can be reduced and

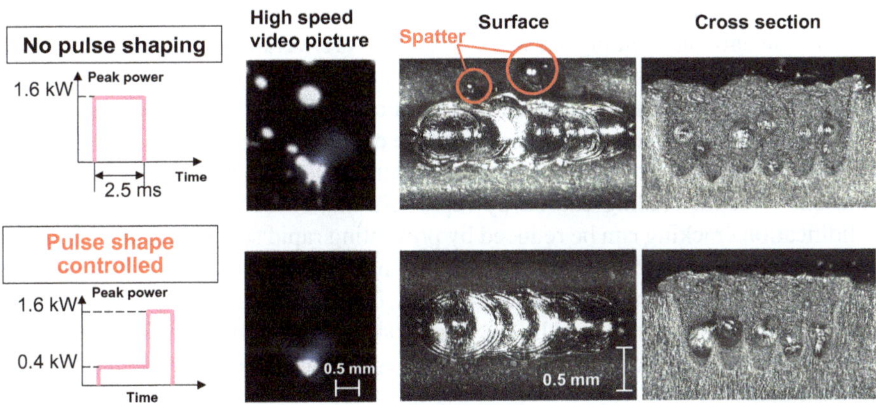

Fig. 4.9 Rectangular and controlled pulse shapes, plume and spattering during spot welding, surface appearances of overlapped spot welds, and cross sections of welds, showing severe spattering and appreciable spatters on surface, but no difference in porosity formation in case of rectangular pulse

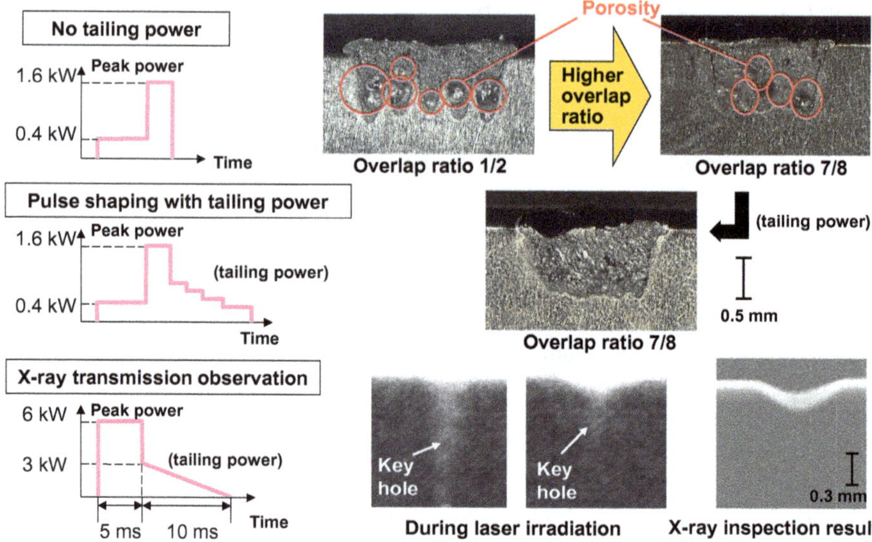

Fig. 4.10 Various pulse shapes, photographs of longitudinal sections of spot seam welds produced with pulsed laser with and without tailing power, X-ray in situ observation results during spot welding and X-ray inspection result, showing effects of overlap ratio and tailing power on porosity reduction

prevented in deep spot welds by selecting power-controllable pulse shape properly. Such results are also obtained in spot welds of aluminum alloys [1, 10].

However, solidification cracks are easily formed in almost all aluminum alloys except for pure aluminum such as A1050 and A1100, Al–Mn alloy A3003, Al–Si alloy A4047. Spot welding results of Al–Mg alloy A5083 are exhibited in Fig. 4.12 [1, 11]. In conventional welding such as TIG or MIG arc welding cracks hardly occur in A5083 alloy. But long solidification cracks are present along grain boundaries in laser spot welds of A5083 alloy sheet subjected to rectangular pulse shape. Besides, the addition of low laser power tailing can reduce cracks drastically. Such solidification cracking is attributed to residual liquid present along grain boundaries in the wide area due to rapid cooling caused by rapid termination of laser power. Therefore, solidification cracking can be reduced by preventing rapid solidification of dendrites and consequently by narrowing the formation area of residual liquid along grain boundaries due to tailing of low power [1, 11]. This effect is confirmed by simulation of molten pool behavior [1, 11]. Laser pulse shaping, which can reduce or prevent spattering, porosity and solidification cracks, is essential to produce sound welds in spot welding.

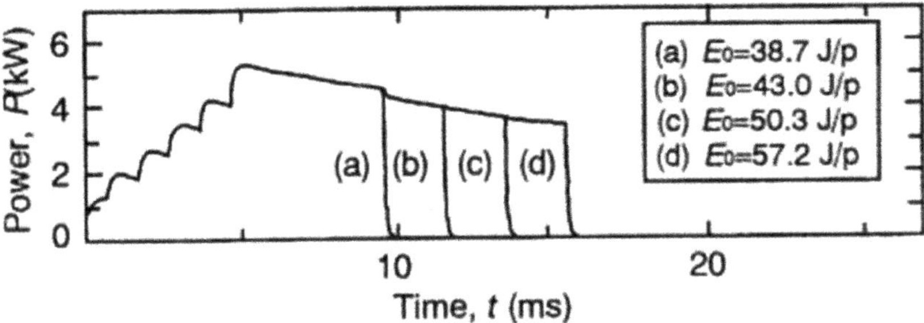

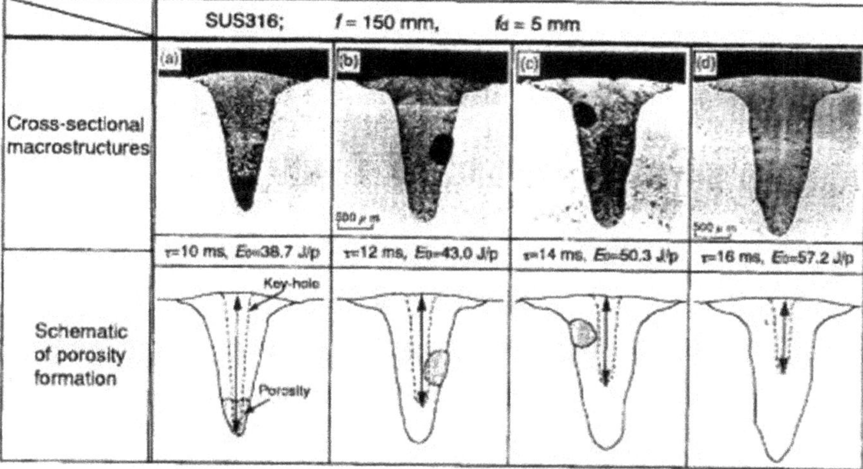

Fig. 4.11 Various laser pulse shapes with tailing powers, and cross sections of Type 316 spot welds made with lasers of various pulse shapes and their schematic spot welds and porosity location, showing effect of laser pulse shaping or falling power on porosity location and disappearance

4.2 Bead Welding with Continuous Wave (CW) Laser

4.2.1 Effect of Laser Welding Parameters on Weld Penetration and Defects Formation

Laser weld beads of various penetration depths can be produced by high-power CW (continuous wave) lasers. The penetration depths depend upon the welding conditions such as laser power, defocused distance, (power density, energy density,) laser wavelength, shielding gas and its flow rate, gas shielding method and welding speed. The effects of gas and flow rate on weld penetration and welding phenomena were described in CO_2 laser welding in 3.2 and then should be referred to the Sect. 3.2. The effect of the defocused distance or laser power density on the weld penetration should be referred to the photographs of cross-sectional weld beads made with YAG

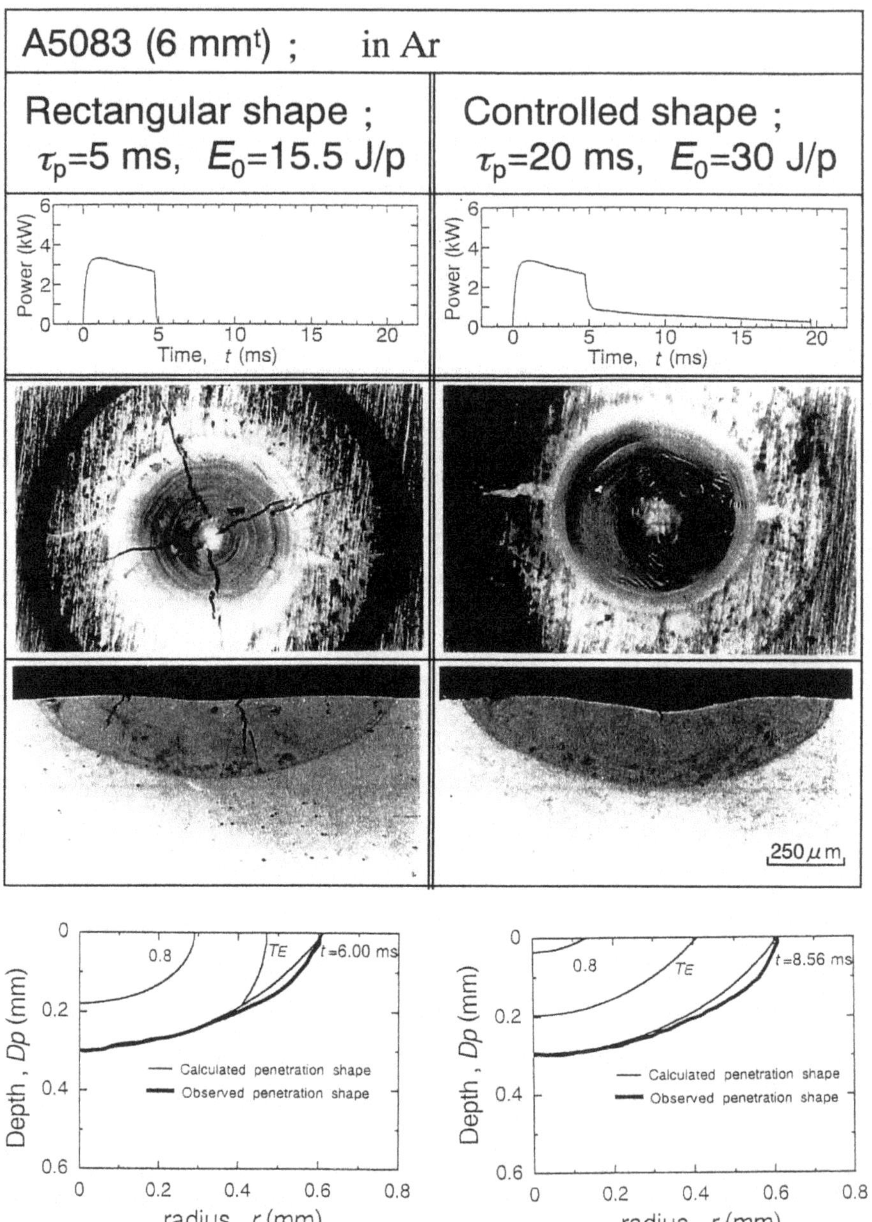

Fig. 4.12 Rectangular pulse shape and controlled shape with tailing power, their surfaces and cross sections of spot welds in Al alloy A5083, and examples of calculated mushy zones during welding with actual weld fusion boundaries, showing effect of tailing power on narrowing of mushy zone and consequent reduction in solidification crack length

and fiber lasers in Fig. 2.4 in Sect. 2.1. The penetration depths become shallower with an increase in the defocused distance, and the welds are changed from a keyhole type to a heat conduction type. If the laser power or power density is high, the deepest weld bead may be obtained at the defocused distance of −4 mm (below the plate surface), for example. The distance depends upon the laser power or power density and is longer with an increase in the power or power density. In this section below, the welding results obtained with a fiber laser or a disk laser are treated, because recently these lasers are predominantly mostly used and the effect of a shielding gas is small in fiber or disk laser welding.

The penetration depths of weld beads were investigated by changing the welding speeds in fiber laser welding of Type 304 steel plates at 6 kW. The surface appearances and the cross-sectional photographs of weld beads are already shown as a function of welding speed in Fig. 3.13. The weld beads are produced by a keyhole type, and the penetration is becoming shallower from 7 to 2.5 mm depth with an increase in the welding speed from 25 to 250 mm/s (15 m/min). Porosity is formed at the low welding speeds of 25 and 50 mm/s (3 m/min), while a large size of spatters are present on the plate surface and consequently underfilled weld beads are produced at the high welding speeds of 150 mm/s (9 m/min) and 250 mm/s.

The effects of fiber laser beam diameter, welding speed and laser power on weld penetration and welding defects formation in Ar shielding gas are summarized in Fig. 4.13 [12, 13]. Type 304 steel plates of 8 and 20 mm in thickness were employed. At 6 kW, the penetration depth decreases with an increase in the welding speed. The full penetration weld beads of an 8-mm-thick plate are obtained at higher welding speeds for smaller beam diameter or higher power density on the focal point. In the

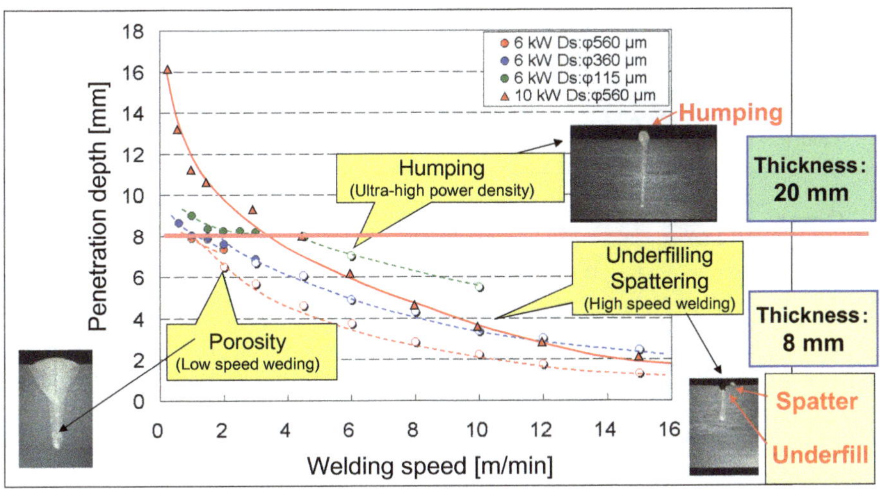

Fig. 4.13 Effects of fiber laser beam diameter, welding speed, and laser power on penetration depths and welding defects formation of weld beads in Type 304 plates of 8 or 20 mm thickness in Ar shielding gas

case of the smallest beam diameter of 115 μm, the penetration depth is the deepest especially at high welding speeds due to the highest power density. At low welding speeds, the weld penetrations are almost equal since the power density at the bottom of a keyhole for the laser of a small beam diameter is not so high. It is clear that the penetrations at high laser power of 10 kW are deeper than that at the powers of 6 kW only at the low welding speeds of less than 3 m/min (50 mm/s). It is understood that higher power CW lasers are required to produce deeper weld beads. In addition, it is suggested from Fig. 4.13, that, in the case of the beam diameter of 360 and 560 μm, porosity is formed at low welding speeds of less than 3 m/min, while underfilled weld beads are produced at high welding speeds of more than 8 m/min. On the other hand, in the case of the smallest beam diameter of 115 mm, underfilled weld beads are not formed but humping is likely to occur at high welding speeds of more than 6 m/min.

The effect of power density on the weld penetration was investigated by using fiber lasers of different beam spot diameters. The surface appearances and cross sections of weld beads made at the power of 10 kW and the welding speed of 4.5 m/min are compared among different diameters leading to different power densities in Fig. 4.14 [12, 13]. Deep partial-penetration weld beads are formed. The penetration is deeper with an increase in the power density caused by smaller spot diameter. In addition,

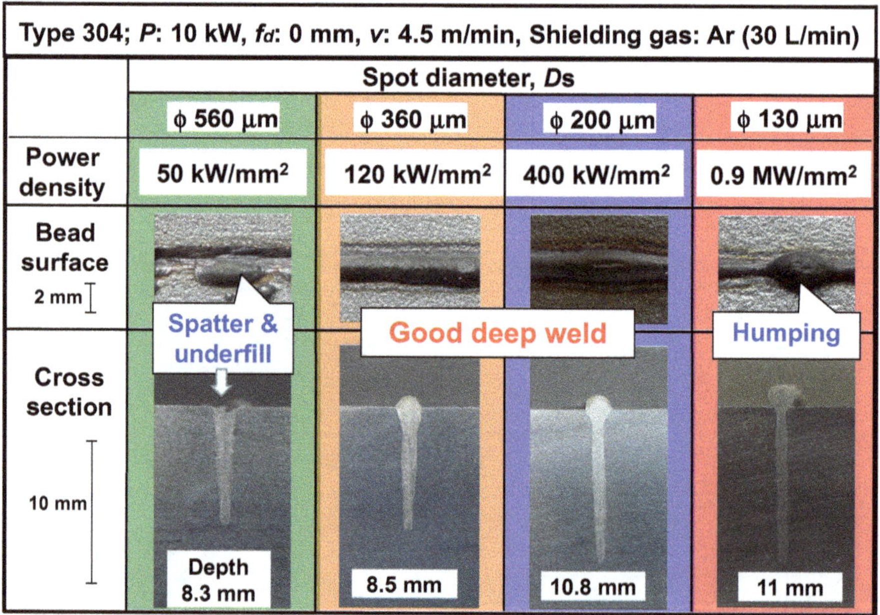

Fig. 4.14 Surface appearances and cross sections of weld beads made with fiber laser of different spot diameters at 10 kW power and 4.5 m/min welding speed, showing effects of spot beam diameter and consequent laser power density at focal point on weld penetration depth and geometry, and welding defects

in the case of 560 μm diameter, spatters are generated, resulting in the formation of an underfilled bead. On the other hand, in the case of 130 μm diameter, a humping bead is observed. Nevertheless, sound deep weld beads are produced in welding with a fiber laser of 200 and 360 μm diameter. In laser welding, sound good weld beads can be made by properly selecting the optimum welding conditions.

4.2.2 Plume Behavior and Physical Phenomena during CW Laser Welding

The welding phenomena such as the behavior and features of a laser-induced plume, the presence and motion of a keyhole, melt flows inside the molten pool, the formation and flows of bubbles leading to pores (or porosity) are observed by high-speed video camera and X-ray transmission method, or are analyzed by spectroscopy. Figure 4.15 shows a schematic representation of simultaneous in situ observation system of behavior of a plume, a keyhole and a molten pool during laser welding [13]. And Fig. 4.16 exhibits an example of simultaneous in situ observation results of a

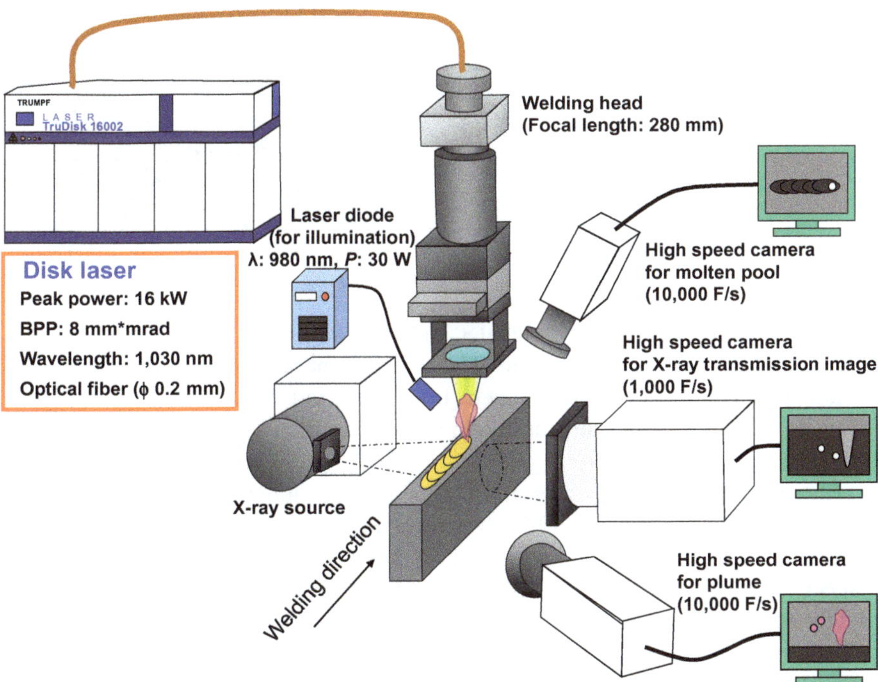

Fig. 4.15 Schematic representation of high-speed video cameras and X-ray transmission imaging system for simultaneous in situ observation of plume behavior, keyhole motion, and molten pool behavior during laser welding

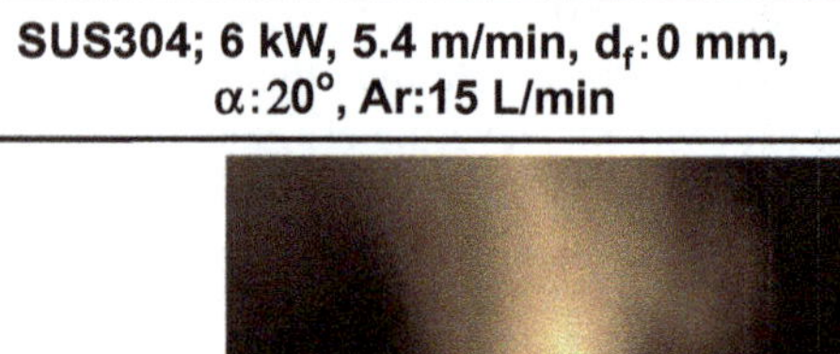

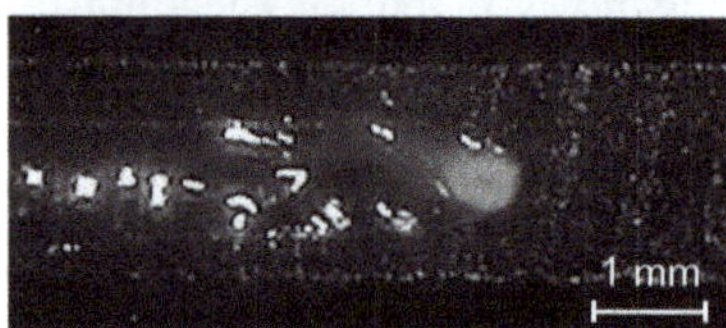

Fig. 4.16 Example of simultaneous in situ observation results of plume, molten pool and keyhole inlet on surface, and keyhole inside molten pool during CW disk laser welding of Type 304 at 6 kW and 5.4 m/min

plume, a molten pool and a keyhole inlet, and a keyhole during CW disk laser welding of Type 304 at 6 kW and 5.4 m/min [13]. A simultaneous behavior observation of a keyhole and a molten pool is feasible by high-speed video cameras at the high speed of 10,000 F/s and X-ray transmission in situ observation imaging at the high speed of 1000 F/s.

The spectroscopic analysis results of laser-induced plumes during welding of Type 304 with 1.5 kW YAG laser of 0.58 mm beam diameter at 20 mm/s, and with 10 kW fiber laser of 0.13 mm beam diameter at 50 mm/s are shown in Figs. 4.17 and 4.18, respectively [14, 15]. In YAG laser welding, Fe(I) peaks meaning neutral atoms are mainly detected, and Fe(II) peaks meaning ions are not detected. The temperature of a plume is estimated to be about 3600 K from Saha's equation, which signifies that the temperature is not high. On the other hand, in the case of extremely high-power density of fiber laser, many peaks of violent and blue light are detected from the plume but the light from Ar(I) is not detected. The temperature of the plume is estimated to be about 6000 K, which should be called a weakly ionized plasma [15].

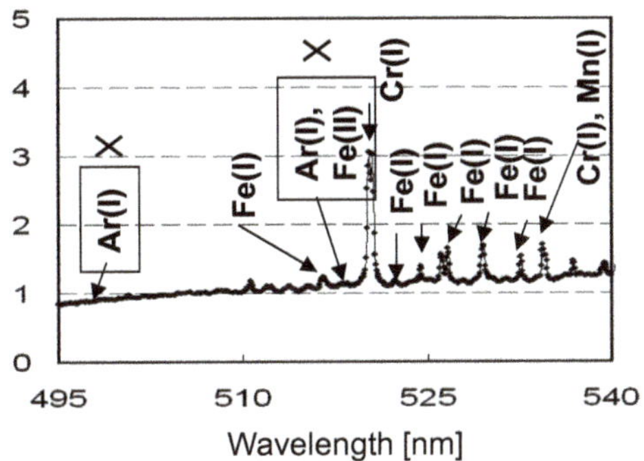

Fig. 4.17 Spectroscopic analysis result of laser-induced plume during welding of Type 304 stainless steel with 1.5 kW YAG laser of 0.58 mm beam diameter at 20 mm/s

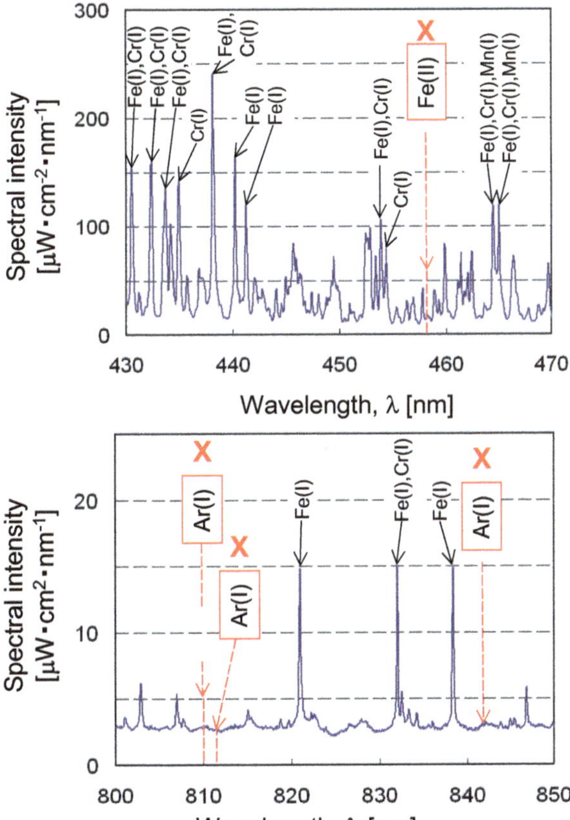

Fig. 4.18 Spectroscopic analysis result of laser-induced plume during welding of Type 304 stainless steel with 10 kW fiber laser of 0.13 mm beam diameter at 50 mm/s

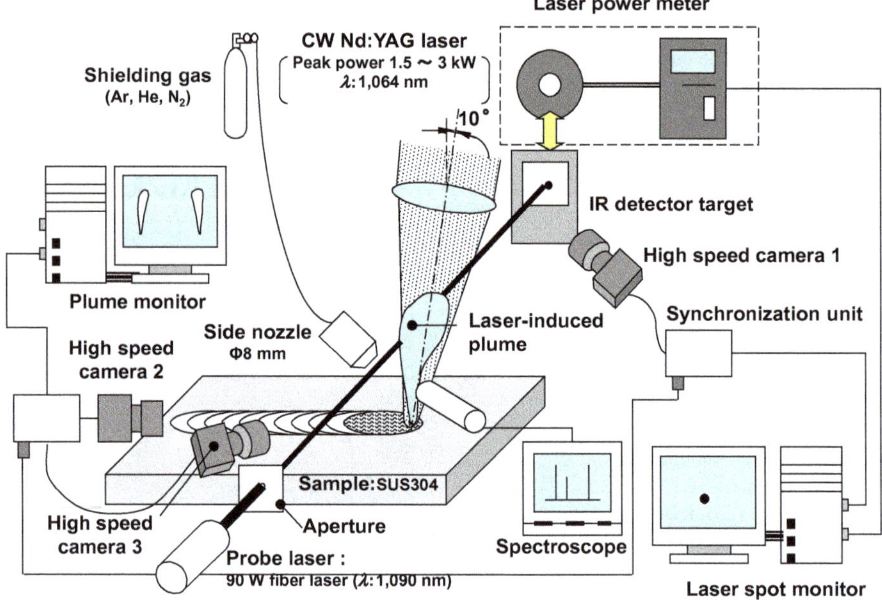

Fig. 4.19 Schematic measurement setup of power attenuation and movement observation of probe laser passing through plume during laser welding

It is concluded from these results that the temperature of the plume during welding with a laser of about 1 μm wavelength is not so high as supposed.

The movement of a probe laser beam due to the interaction between a plume and ultrafine particles (corresponding to high temperature range) and a probe laser beam was observed through high speed video, and the attenuation of a probe laser such as fiber laser, diode laser, or He–Ne laser through the plume zone was measured, as shown in Fig. 4.19 [14]. The attenuations of respective lasers are indicated in Fig. 4.20 [14]. The attenuations values are in inverse proportion to 4th power of laser wavelength. This result suggests that Rayleigh scattering occurs due to ultrafine particles, etc. during laser welding and that the effect of Rayleigh scattering on the laser attenuation is greater for a laser with shorter wavelength. Besides, the effect is confirmed to be very small in fiber laser welding in a shielding gas.

The effect of a tall plume on weld penetration was also confirmed by remote welding with a CW fiber laser without fan or gas flow or with fan. Figure 4.21 shows both the top and bottom surfaces and cross sections of laser melt run weld beads made in Zn-coated steel sheet of 1.5 mm thickness and bare steel sheet of 1.2 mm thickness at the focal point without fan in air atmosphere [16]. The full deep penetration welds change into the partial shallow penetration welds on the way of welding. Figure 4.22 exhibits the bottom surfaces of laser weld beads made, and examples of plumes and molten pools during laser welding with fan and without fan in air environment [16]. The effect of a tall plume (a long high-temperature zone with low density) on

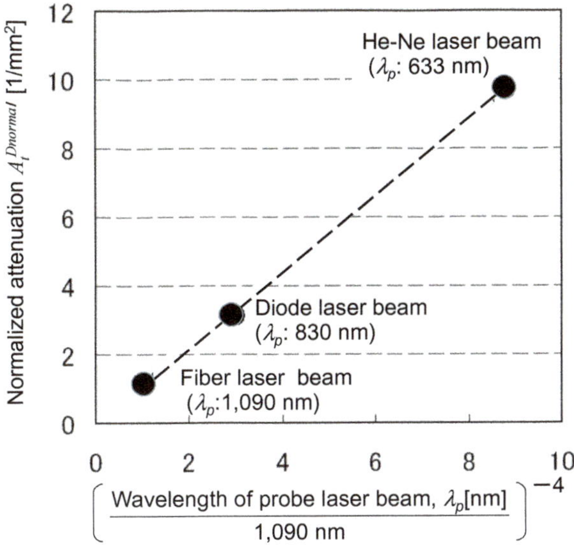

Fig. 4.20 Attenuation of fiber laser, diode laser, and He–Ne laser after passing through plume, indicating feasibility of Rayleigh scattering in inverse proportion to 4th power of laser wavelength

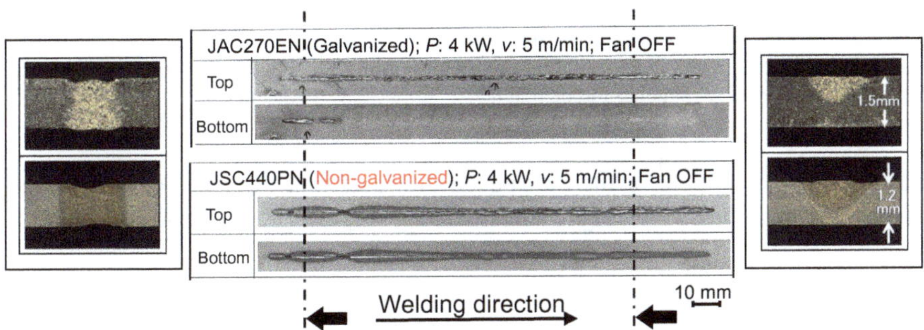

Fig. 4.21 Top and bottom surfaces and cross sections of laser melt run weld beads made in Zn-coated steel sheet of 1.5 mm thickness and bare steel sheet of 1.2 mm thickness at focal point without fan in air atmosphere, showing change in weld penetration during laser welding

phenomena and weld penetration is clear. The penetration becomes shallower and a keyhole type of a deep weld changes into a partially penetrated weld because a plume grows taller during laser welding without fan, although a keyhole type of a deep weld is constantly made by forming a short plume or high-temperature, low-density zone in welding with fan. A low-density zone is confirmed to grow together with high-temperature vapors from the keyhole above the sheet. The effect of a tall plume on the weld penetration is interpreted by considering the effects of a low-density zone shifting the focal point of focusing lenses downward and refracting an incident laser beam. From such observation and analytical results, the effects of plume behavior

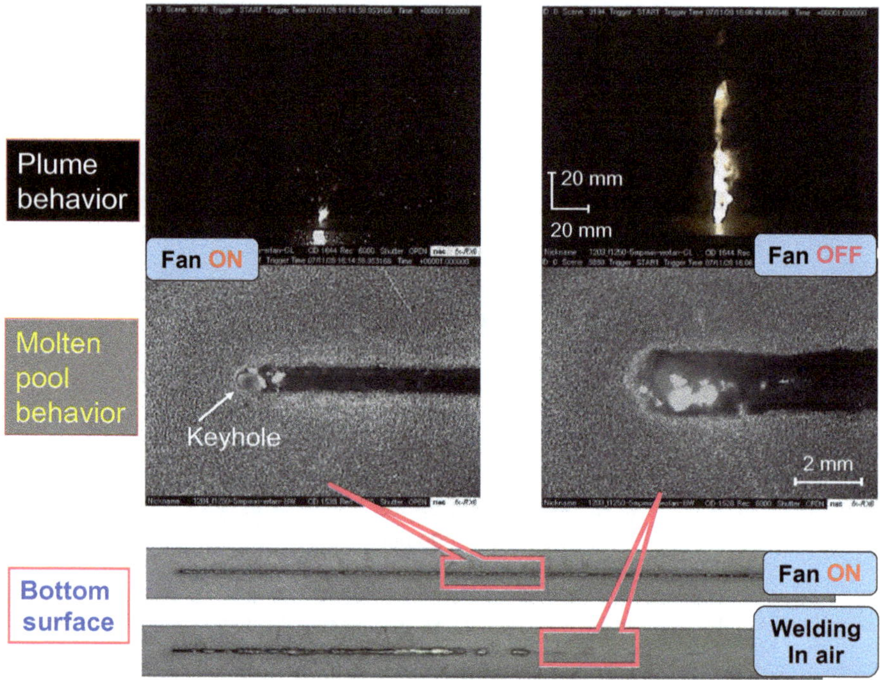

Fig. 4.22 Bottom surfaces of laser weld beads made, and examples of plumes and molten pools during laser welding with fan and without fan in air environment, showing suppression of variation in penetration depth by reducing plume height

and physical phenomena on the weld penetration and welding results in laser welding can be understood by referring to schematic illustration already shown in Fig. 2.10.

4.2.3 *Melt Flows and Spattering during Laser Welding*

Melt flows in the molten pool during laser welding affect the weld penetration and geometry. The weld beads produced in Type 304 steel plate with YAG laser of 3.3 kW at 10 mm/s in Ar shielding gas in Ar atmosphere and in air atmosphere, and the melt flows observed during the laser welding and their schematic melt flows near keyhole inlet are comparatively shown in Fig. 4.23 [17, 18]. ZrO_2 powders were used as tracers of melt flows. In Ar atmosphere with negligibly low oxygen, a wider surface bead is formed while in air atmosphere a weld bead of narrower surface is produced. It was conventionally insisted that the formation of a wide surface bead was due to the thermal radiation of a high-temperature plasma ejected from the keyhole inlet. However, in fact, in YAG laser welding, a plume (of weakly ionized plasma) is ejected from the keyhole inlet and its temperature is not so high as conventional supposition.

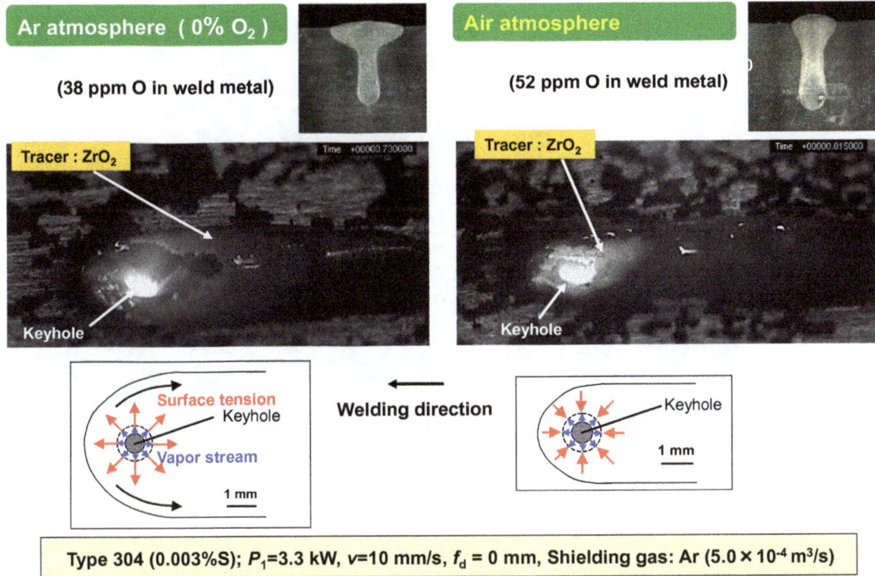

Fig. 4.23 Weld beads made by YAG laser welding of Type 304 steel plate at 3.3 kW and 10 mm/s under Ar shielding gas flow in Ar atmosphere and in air atmosphere, and high-speed video observation results and schematic illustration of melt flows on molten pool surface during welding, showing different weld bead geometries affected by melt flows and effect of surface tension on melt flows near molten pool surface

Consequently, a difference in the surface bead width should be interpreted by the differences in the surface tension and the consequent melt flows near the molten pool surface. That is to say, the weld geometry with a wide surface bead is attributed to the melt flows from a keyhole inlet to the lateral and rear peripheral molten pool because the surface tension is higher at lower temperatures in Ar atmosphere, while the narrower weld bead surface is due to the melt flows approaching to the keyhole inlet near the molten pool surface because the surface tension is changed to be higher at higher temperatures due to the oxygen content absorbed in the molten pool in air atmosphere. Moreover, the melt downward flows along the keyhole wall are observed in the molten pool in air atmosphere, leading to slightly deeper weld penetration. In conclusion, it is confirmed in laser welding of steels at low welding speeds in Ar gas shielding that the weld bead penetration and geometry are affected by the surface tension and the melt flows near the molten pool surface according to the effect of oxygen in Ar or air atmosphere.

Melt flows near the molten pool surface were observed during fiber laser welding of Type 304 steel plate at 6 kW and 3 m/min through two high-speed video cameras. As the results are shown together with the surface and cross section of a weld bead and X-ray inspection result in Fig. 4.24 [18]. Although the melt overflows toward the rear molten pool from the keyhole inlet, a sound good-looking weld bead can be produced. Figure 4.25 shows photographs taken by high-speed video cameras during

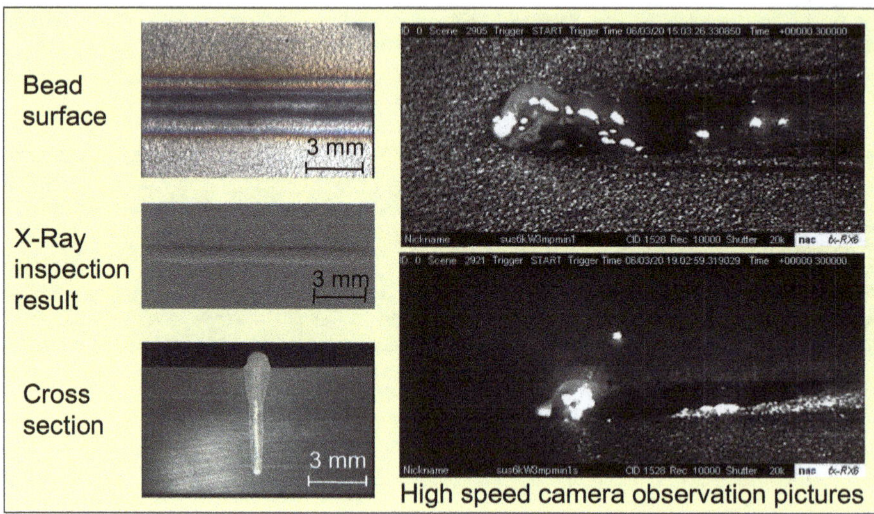

Fig. 4.24 Surface appearance, X-ray inspection result, and cross section of weld bead made by fiber laser welding of Type 304 steel plate at 6 kW and 3 m/min, and melt flows near molten pool surface observed through two high-speed video cameras during laser welding

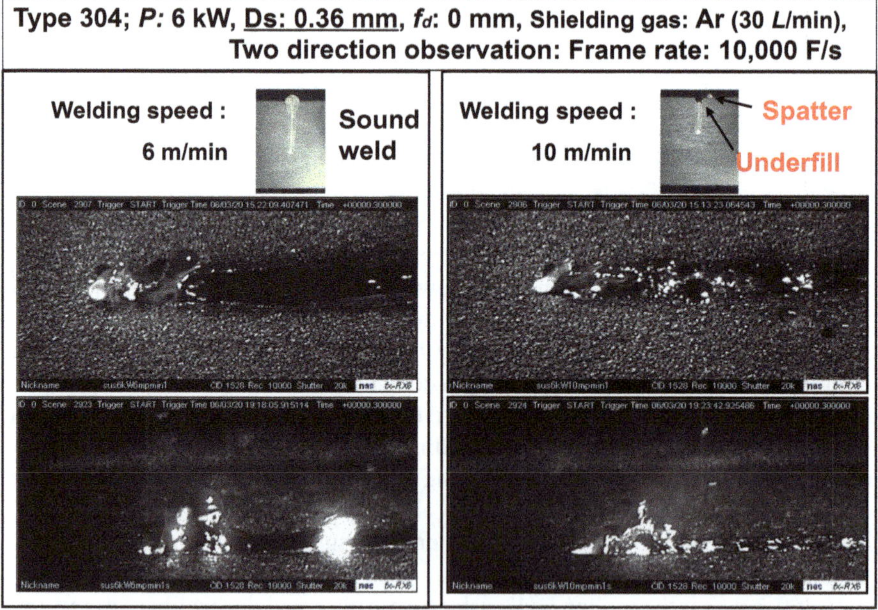

Fig. 4.25 High-speed video photographs of molten pools during welding of Type 304 with fiber laser of 0.36 mm in spot beam diameter at 6 and 10 m/min

welding with a fiber laser of 0.36 mm in spot beam diameter at 6 and 10 m/min [19]. At 6 m/min, the melt overflows in the rear molten pool, resulting in a sound weld bead. At 10 m/min, however, the melt overflows behind the molten pool, leading to an underfilled weld bead.

Melt flows in the molten pool during laser welding with a fiber or disk laser were also observed by the transmission in situ observation system of two sets of microfocused and mini-focused X-rays, using WC particles of about 0.5 mm diameter as tracers. This system is schematically shown in Fig. 4.26 [20]. Examples of photographs showing keyhole and WC particles set about 2, 3, and 5 mm below the plate surface and measurement 3D movement of the WC particles are shown in Fig. 4.27 [20]. Melt flows in the molten pool during welding at 25, 100, and 150 mm/s are exhibited from such measurement results using many WC particles in Fig. 4.28 [20]. There are two main different melt flows. One is the circular flows backward along the molten pool bottom from the keyhole tip to the rear bottom pool and then forward in the middle part or the upper part from the rear pool to the keyhole. The other is backward melt flows including partly circular flows in the upper molten pool from the keyhole inlet to the rear molten pool. At the high speeds of 100 and 150 mm/s, the rear wall melts of the keyholes are ejected as large spatters over the molten pools probably caused by strong shear streams of evaporated matters called vapors or plumes. Consequently, underfilled (concave) weld beads are formed. Underfilling can be reduced by decreasing severe spattering. The mechanism and the

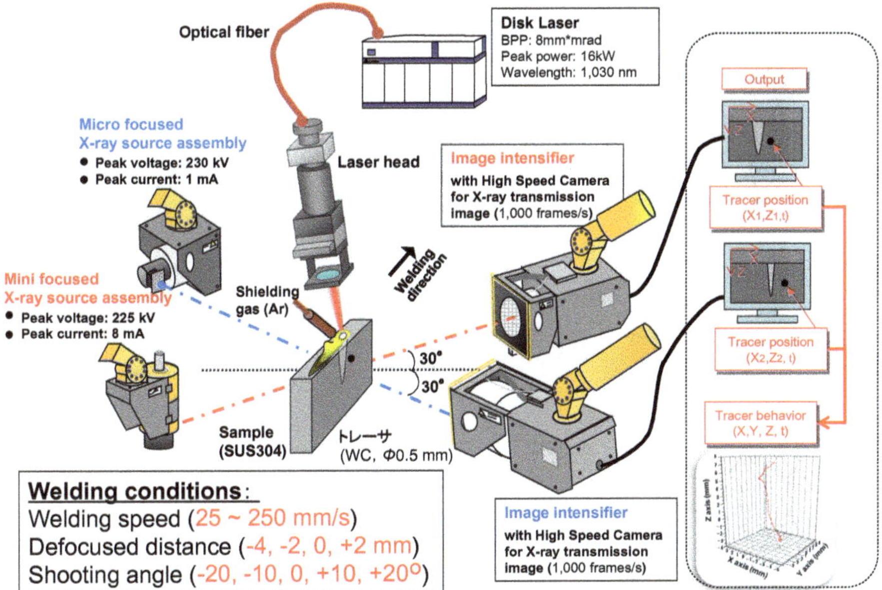

Fig. 4.26 Schematic representation of transmission in situ observation system of two sets of micro-focused and mini-focused X-rays, using WC particles of about 0.5 mm diameter as tracer of melt flows in molten pool

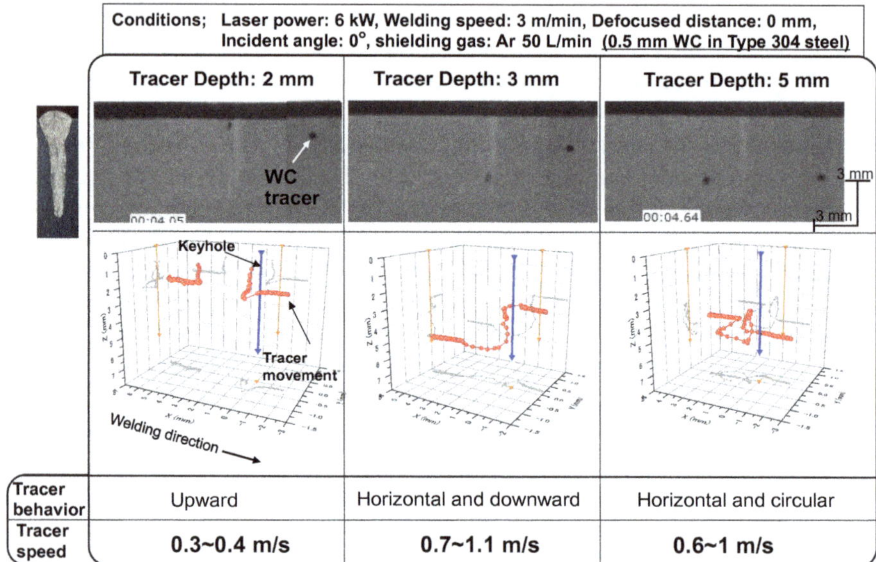

Fig. 4.27 X-ray transmission observation results of keyhole and WC particle set about 2, 3, and 5 mm below plate surface, measurement of 3D movement of WC particle, tracer behavior and tracer speed in molten pool

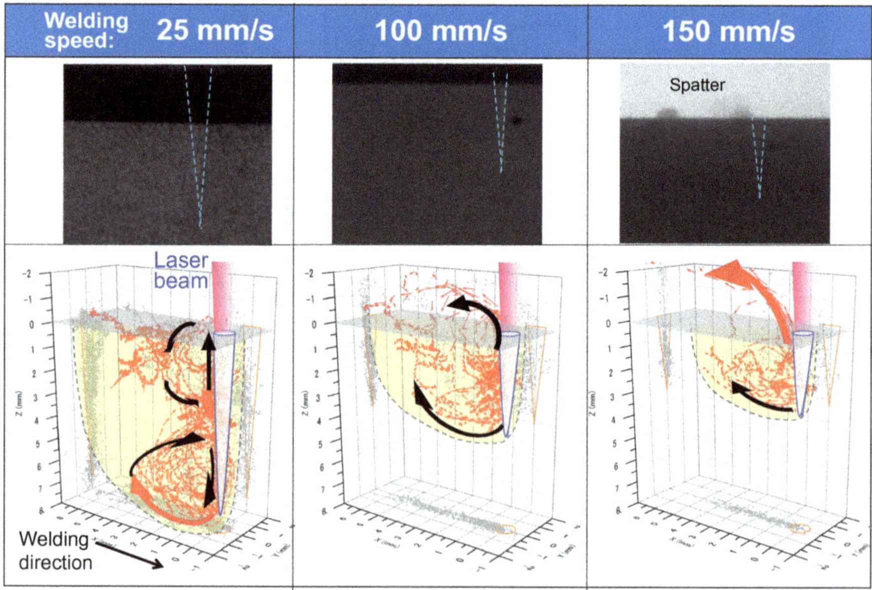

Fig. 4.28 X-ray observation results and melt flows (obtained by movement of many particles) in molten pool during laser welding at 25, 100, and 150 mm/s. showing two circulated melt flows, and fast bottom flow at low welding speeds and fast upper flow from keyhole inlet at high welding speeds

preventive procedures and welding conditions of spattering leading to underfilling will be described in detail in Sect. 5.4.

4.3 Modeling and Simulation of Laser Welding

4.3.1 Properties of Plume and Plasma

Plasma diagnostics using spectroscopy is one of the beneficial procedures for understanding of the interaction between a laser beam and plume or plasma during laser welding. The profile and intensity of emission spectrum of atoms are important information sources.

In the temperature measurement of a plasma [21], when the plasma is in the state of local thermal equilibrium (LTE), the Boltzmann distribution law is established, and thus, the atomic number N_n excited up to the energy level n is expressed in the following equation:

$$N_n = N(g_n/Z)\exp(-E_n/kT) \tag{4.1}$$

Here,

N total atomic number per unit volume of plasma
g_{n} statistical weight of excited n level
E_{n} excitation energy of excited n level
k Boltzmann constant.

Spectrum line intensity of a plasma is proportional to atom concentration (or ion concentration) in the excited state given from the Boltzmann distribution, and thus, the excitation temperature of atoms (or ions) is obtained by measuring the spectrum line intensity. When the spectrum line of frequency ν_{nm} is emitted at transition from the excited level n to the level m, the spectrum line intensity $\mathrm{I_{nm}}$ is expressed in the following equation:

$$I_{nm} = N(g_n/Z(\mathrm{T}))A_{nm}h\upsilon_{nm}\exp(-E_n/kT) \tag{4.2}$$

Here,

$Z(T)$ partition function of atoms at temperature T
A_{nm} transition probability of n to m
h Planck's constant.

When the logarithm is taken and λ_{nm} is rewritten from ν_{nm}, the following equation is obtained:

$$\mathrm{In}(I_{nm}\lambda_{nm}/g_nA_{nm}) = -E_n/kT + \mathrm{C} = -(5040/T)E_n + \mathrm{C} \tag{4.3}$$

Here,

C constant.

When In (I_{nm} λ_{nm}/g_n A_{nm}) and E_n (eV) are set at the vertical axis and the horizontal axis, respectively, and their relationships for various spectrum lines are plotted, one straight line is obtained. Then, the excitation temperature is calculated from the gradient of the line. Since the values of the transition probability differ in the literature, the excitation temperature obtained may be different depending upon the literature. If one straight line is not obtained, it is judged that the plasma should not be in the state of the local thermal equilibrium (LTE).

Electron density can be obtained by Saha's ionization equation or Shark broadening.

The ionization constant S_j of the component j is expressed according to Saha's ionization equation as follows:

$$S_j = n_i n_e / n_a = \left((2\pi mkT)^{3/2}/h^3\right)(2Z_i/Z_a)\exp(-V/kT) \tag{4.4}$$

Here,

n_a number density (concentration) of atoms of component j
n_i number density (concentration) of ions of component j
n_e electron density
V ionization energy.

n_e is readjusted from Saha's equation based upon the relationship between the densities of atoms and ions and spectrum line intensity and the partition function, and then the following equation is obtained.

$$n_e = (I_a/I_i)(g_i A_i v_i / g_a A_a v_a)\left(2(2\pi mkT)^{3/2}/h^3\right) exp((E_a - E_i - V)/kT) \tag{4.5}$$

When the temperature and the relative intensity of atom lines and ion lines of the same element, the electron density of a thermal equilibrium plasma is obtained from the above equation.

At the laser power density of 10^9 to 10^{12} W/cm^2, physical phenomenon is controlled by the process called the inverse Bremsstrahlung. Photons are absorbed by the interaction of ions and electrons. The intensity of photons passing through a plasma and absorption coefficient, α, due to the inverse Bremsstrahlung are expressed in the following:

$$I = I_0 \exp(-\alpha x) \tag{4.6}$$

$$\alpha = n_i n_e Z^2 e^6 \lambda^2 In(2.25kT_e\lambda/hc)/24\pi^3 \varepsilon_0^2 c^3 m_e k T_e (2\pi\ m_e k T_e)^{1/2} \tag{4.7}$$

According to this process, the absorption coefficient increases proportional to the square of the laser wavelength. Therefore, the interaction of CO_2 laser is 100 times larger than that of YAG laser, disk laser or fiber laser since the wavelength of CO_2 laser is about ten times longer than that of YAG laser, disk laser, or fiber laser.

4.3.2 Rise in Material Temperature

When the laser is irradiated on the material, free electrons in it must be vibrated according to the laser frequency, and the laser energy is absorbed. The laser penetration depth (*d*) in aluminum plate is calculated to be about 10 nm for YAG, disk or fiber laser of about 1.03–1.08 μm wavelength and around 12 nm for CO_2 laser of 10.6 μm wavelength. The laser absorption depth in the metal is less than 1 μm and extremely shallow due to the high frequency of laser. Consequently, it is understood that laser heating occurs only on the top surface layer. This heat is transmitted into the material by heat conduction.

When the pulsed laser of constant power density is irradiated on the surface of semi-infinite metal plate, the material temperature of the depth *x* and the surface (*x* = 0) during heating is given in the following equation:

$$T(x,t) - \frac{AP_{\mathrm{d}}}{\lambda}\left[\left(\frac{4\alpha d}{\Pi}\right)^{1/2} \mathrm{e}^{-[\Pi^{f}/(4\alpha)^{1/2}]^2} - xerfc\left(\frac{x}{(4\alpha d)^{1/2}}\right)\right] + T_{*} \quad (4.8)$$

$$T(0,t) = \frac{2AP_{\mathrm{d}}}{\lambda}\sqrt{\frac{\alpha t}{\Pi}} + T_0 = 2AP_{\mathrm{d}}\sqrt{\frac{t}{c\rho\lambda\Pi}} + T_0 = \frac{2AP_{\mathrm{A}}}{c\rho}\sqrt{\frac{t}{\Pi\alpha}} + T_{\mathrm{e}} \quad (4.9)$$

Here,

- A laser absorptivity (=1 − R)
- λ thermal conductivity
- t time
- α thermal diffusivity (= $\lambda/(c\rho)$)
- c specific heat
- ρ density.

Besides, when bead welding is performed at the speed of *v* in the *x*-direction by continuously irradiating a CW laser of the power density of P_{d} (*x*, *y*) on the material, the temperature at the point (*x*, *y*, *z*) in the steady state without radiation loss from the surface is given in the following equation:

$$T(x,y,z,t) = T_0 + \frac{A}{4c\rho(\Pi\alpha)^{3/2}} \cdot$$
$$\int_0^{\infty}\int_{-\infty}^{\infty}\int_{-\infty}^{\infty} \frac{P_d(x',y')}{\sqrt{t^3}} \exp\left[-\frac{(x-x'+vt)^2+(y-y')^2+z^2}{4\alpha t}\right] \mathrm{d}x'\mathrm{d}v'\mathrm{d} \quad (4.10)$$

In order to raise the material temperature, T, quickly and melt the material, higher laser absorptivity, A, of the material and/or a higher laser power density is needed. The temperature rise is slow in the case of the material with high thermal diffusivity. Actually, a laser beam of low power density is more difficult to melt than steel. This is attributed to higher laser reflectivity and higher thermal diffusivity (or higher thermal conductivity) of aluminum alloy. On the other hand, in the case of a laser of high power and high power density, a keyhole type of welding is performed due to the intense recoil pressure of severe metal evaporation. Consequently, it is sometimes obtained that the laser weld penetrations of aluminum alloys are deeper than those of steels. It signifies that a keyhole type of laser weld beads are predominantly influenced and governed by evaporation.

4.3.3 Simulation during Laser Welding

Recently, laser welding results and phenomena can be understood by observation using high-speed video cameras, X-ray transmission in situ observation method, spectroscopic measurement, and so on. Especially, laser welding phenomena such as plume and keyhole behavior, and molten pool motion are very unstable, and thus the modeling in quasi-steady-state cannot display the actual welding phenomena and the establishment of dynamic modeling has been required. Recently, good modeling and simulation of laser and hybrid welding have been developed and demonstrated by Prof. S-J. Na's group (KAIST, South Korea), Prof. A. Otto (Vienna University of Technology, Austria), Mr. T. Nakagawa (Panasonic Welding System, Japan), and so on. An example of the results of Prof. Na and Dr. Cho is shown in Fig. 4.29 [22]. It appears that the dynamic motion displayed is similar to the actual laser welding phenomena.

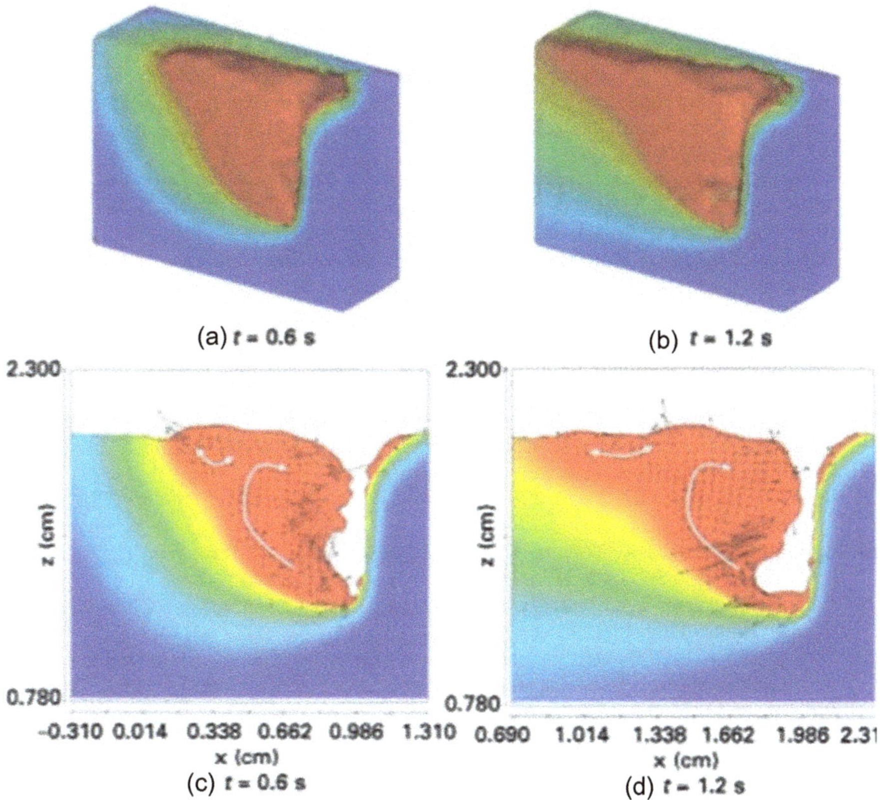

Fig. 4.29 Calculated temperature profiles and flow patterns in three-dimensional cross-sectional view **a**, **b**, and cross-sectional side view **c**, **d** during laser welding

References

1. Kim J-D (1997) Study of temporal and spatial high-resolution measurement, and formation mechanisms and preventive procedures of defects. Thesis for Doctor of Engineering (in Japanese)
2. Matsunawa A, Kim J-D, Katayama S, Semak VV (1996) Proceedings of ICALEO 96, LIA, Detroit, USA, vol 81, pp 58–67
3. Matsunawa A, Kim J-D, Takemoto T, Katayama S (1995) Proceedings of ICALEO 95, LIA, California, USA, vol 80, pp 719–728
4. Kim J-D, Katayama S, Mizutani M, Takemoto T, Matsunawa A (1996) The review of laser engineering. Laser Soc Jpn 24(9):996–1005 (in Japanese)
5. Matsunawa A, Yoshida H, Katayama S (1984) Proceedings of ICALEO '84, LIA, USA, vol 44, pp 35–42
6. Katayama S, Kohsaka S, Mizutani M, Nishizawa K, Matsunawa A (1993) Proceedings of ICALEO '93, LIA, pp 487–497
7. Katayama S, Seto N, Mizutani M, Matsunawa A (2001) Proceedings of ICALEO 2001, LIA, Session C: Welding, 804 (CD), p 804

8. Katayama S, Matsunawa A, Morimoto A, Ishomoto S, Arata Y (1983) Proceedings of ICALEO '83, LIA, pp 127–134
9. Nakamura H (2015) Elucidation of fundamental phenomena in laser welding of titanium, and evolution to precision micro-joining. Thesis for Doctor of Engineering (in Japanese)
10. Katayama S, Mizutani M, Ikeda H, Nishizawa K, Matsunawa A (1992) Proceedings of ICALEO '92, LIA, pp 547–556
11. Katayama S, Matsunawa A (1998) Proc CISFFEL 6–1:215–222
12. Katayama S, Kawahito Y, Kinoshita K, Matsumoto N, Mizutani M (2007) Proceedings of ICALEO 2007, LIA, Florida, USA, vol 100, pp 353–359
13. Kawahito Y, Mizutani M, Katayama S (2009) Sci Technol Weld Join 14(4):288–294
14. Kawahito Y, Kinoshita K, Katayama S, Tsubota S, Ishide T (2006) J Laser Mater Proces JLPS 13(1):41–47 (in Japanese)
15. Kawahito Y, Kinoshita K, Matsumoto N, Mizutani M, Katayama S (2017) J Japan Weld Soc JWS 25(3):455–460 (in Japanese)
16. Oiwa S, Kawahito Y, Mizutani M, Katayama S (2011) J Laser Appl 23(2):022007-1-7
17. Katayama S, Kawahito Y (2007) J High Temp Soc Jpn, HTSJ 33(3):118–127 (in Japanese)
18. Y. Naito, M. Mizutani and S. Katayama: J. of the Japan Welding Society, JWS, Vol. 24, No. 2 (2006), pp. 149–161. (in Japanese)
19. Kinoshita K, Mizutani M, Kawahito Y, Katayama S (2007) J Jpn Weld Soc JWS 25(1):18–23 (in Japanese)
20. Kawahito Y, Nakada K, Uemura Y, Mizutani M, Nishimoto K, Kawakami H, Katayama S (2016) J Jpn Weld Soc JWS 34(4):239–248 (in Japanese)
21. Daidoji H, Nakahara T (eds) (1989) Atomic spectrum-measurement and its application-Gakkai-shypan (Society Publishing) Center
22. Na S-J, Cho W-I (2013) Developments in modelling and simulation of laser and hybrid laser welding. In: Katayama S (ed) Handbook of laser welding technologies. Woodhead Publishing Limited, pp 522–552

Chapter 5
Formation Mechanisms and Preventive Procedures of Laser Welding Defects

5.1 Features of Various Welding Defects

The applications of laser welding in many industrial fields are increasing together with the development of laser apparatuses and their peripheral devices. However, welding imperfections or defects leading to a fracture or a disaster of manufactured goods or constructions may take place under improper conditions. Therefore, in order to manufacture sound goods or constructions of high-quality and high reliability with a laser beam, it is inevitably important to understand welding phenomena, and formation conditions and mechanisms of defects, to establish preventive procedures for welding defects, and to take the necessary measures for no welding imperfections and defects, and good mechanical properties.

In laser welding, various welding imperfections or defects such as poor welds due to unevenness or gap, deformation sometimes called distortion, strain or warp, underfilling, rough surface or pit formation due to spattering or unstable keyhole, hot cracking and porosity may occur depending upon the kinds of materials and their compositions, welding conditions, and so on. These kinds, occurrence causes, and reduction and preventive procedures are summarized in Table 5.1 [1, 2]. Laser welding defects are classified into three characteristic groups: geometrical or appearance defects, internal or invisible defects, and property or quality defects, as well as normal arc welding defects. These formation mechanisms and importance are sometimes different between laser and arc welding defects. Important laser welding defects to be solved are internal defects called porosity (blowholes, pores, pits, wormhole) and hot cracking (solidification cracking in weld metals and liquation cracking in HAZ (heat-affected zone)). Sound good-looking laser welds with high mechanical properties may be required by reducing spattering. Important defects will be described in detail in the following sections.

S. Katayama, *Fundamentals and Details of Laser Welding*,
Topics in Mining, Metallurgy and Materials Engineering,
https://doi.org/10.1007/978-981-15-7933-2_5

5.2 Formation Mechanism and Preventive Procedures of Porosity

Porosity sometimes refers to a blowhole, a pore, a wormhole, a pit or a bubble. In this book, porosity and bubbles are chiefly used in the "solid" weld fusion zone or weld bead and the "molten" pool, respectively. Porosity is very easily formed during welding with any lasers. The formation mechanisms of porosity can be understood, and accordingly, their preventive measures can be taken by considering three causes separately. One is the formation of bubbles due to an unstable keyhole during keyhole

Table 5.1 Kinds, features, schematic representation, causes, and suppression or preventive procedures of main laser welding defects

Features	Kinds	Depiction	● Causes and O Remedy for prevention
Geometrical defect	◎ Deformation (Strain) (Distortion)		● Great thermal effect ● Incomplete Jigging and restraint conditions ○ High-speed, deep-penetration welding ○ Utilization of rigid jig
	◎ Bad appearance (Dirty surface) (Rough surface)		● Improper welding conditions ○ Low power density welding ○ Removal (Blow away) of evaporated particles
	◎ Drop-through (Bum-through)		● Thin sheet ● Wide bead width due to excessive energy ○ Optimization of welding conditions (Lower heat)
	◎ Undercut		● Slow welding speed and high heat input ● Properties of weld fusion zone (viscosity) ○ Proper welding conditions (gas, etc.)
	◎ Underfill		● Spattering due to excessively power density and small melt zone ○ Avoidance of low energy and focused conditions ○ Optimization of pulsed laser welding

(continued)

Table 5.1 (continued)

Features	Kinds	Depiction	● Causes and O Remedy for prevention
Internal defect	◎ Crack (Hot crack) (Solidification crack) (Liquation crack) (Crater crack) (Bead crack) (Cold crack)	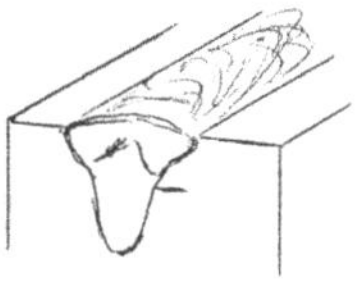	● Low melting point products and segregation ● High stress/strain (rapid strain rate) ○ Proper selection of alloying compositions in weld metal and base metal ○ Optimization of pulsed laser welding ○ Avoidance of high-speed welding ○ Prevention of rapidly augmented strain rate
	◎ Porosity (Blowhole) (Pore) (Pit)		○ Elimination of hydrogen and oxygen sources ○ Formation of stable keyhole ○ Proper selection of shielding gas and flow rate
	◎ Swell in lap joint	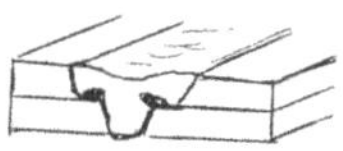	● High-power welding of sheets (deformation) ○ Close adhesion (No gap in lap part) ○ High-speed welding at low heat input ○ Utilization of pulsed laser welding
	◎ Incomplete penetration ◎ Lack of fusion	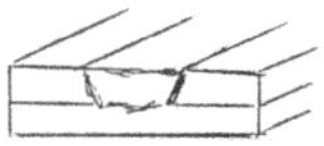	● High reflectivity and effect of plasma ○ Proper conditions for deep penetration ○ Prevention of misalignment
	◎ Inclusion (Oxides, etc.)		○ Surface polishing ○ Proper shielding conditions
	◎ Evaporation loss of alloying elements	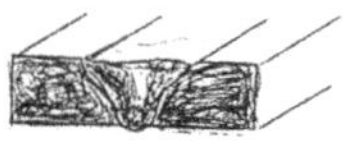	● Volatile elements and low temperature of melt ○ High-speed, deep-penetration welding ○ Pulsed deep-penetration welding

(continued)

Table 5.1 (continued)

Features	Kinds	Depiction	● Causes and O Remedy for prevention
	◎ Macro-segregation		● Incomplete mixing of dissimilar materials ○ Mixing due to beam oscillation or pulsation
	◎ Microsegregation	Cellular dendrite	● Elements of low distribution coefficients ○ Welding at high speed and low heat input ○ Solution heat treatment after welding
Property defect	◎ Mechanical properties	σ (MPa) → Welded joint Base metal ε (%) →	● Formation of soft zones (disappearance of age- or stain-hardening) and defects ○ High-speed, deep-penetration welding ○ Welding conditions for prevention of defects ○ Optimum heat treatment
	◎ Chemical Properties		● Stress corrosion cracking ● Grain boundary corrosion due to precipitates ○ Low heat-input (deep-penetration) welding ○ Relief and relaxation of residual stress

welding, another is the quality of surfaces of sheets or plates in butt- or lap-joint welding, and the other is the material kinds such as Zn-coated or ZAM-coated steels and die-cast alloys. Bubbles generated from a keyhole, resulting in porosity, are characterized by laser welding.

In spot welding with pulsed laser, a deep keyhole is formed, then the laser power is downed quickly, and consequently, porosity is formed from a part of a deep keyhole. The cross sections of YAG laser spot welds and schematic behavior of a keyhole leading to the porosity formation are already shown in Fig. 4.8 [3]. In the lower part of Fig. 4.8, (root) porosity is formed. This is interpreted by the event that the laser power is terminated suddenly when a deep keyhole is formed. When the laser power falls gradually, the formation location of porosity or a pore should move in the upper part of the weld fusion zone. As already shown in Fig. 4.6, the porosity could be avoided when keyholes were shallow in slightly large spot welds. Welds of less than 0.8 mm depth were sound in the case of rectangular pulse shape of about 10 ms width. Sound spot welds of about 1.5 mm depth without porosity may be produced with the pulsed laser of less than 20 ms pulse width under the pulse shape-controlled conditions [4]. It is particularly important to gradually reduce the laser power so as to prevent rapid collapse of a keyhole, as already shown in Figs. 4.10 and 4.11 [4–6].

In bead welding of various metals and alloys with high-power CW laser, porosity or pores are present in their weld fusion zones. The microfocused X-ray transmission observation was performed during bead welding with high-power CO_2 laser, YAG laser, disk laser, and fiber laser to elucidate keyhole behavior, the formation mechanism of bubbles and porosity and melt flows in the molten pool. Examples of cross sections of CO_2 laser welds of A5083 aluminum alloy and Type 304 stainless steel, and X-ray transmission observation photographs and bubble movement during welding are shown in Fig. 5.1 [1, 7, 8]. It is clearly seen that small and large bubbles are mainly generated from the keyhole tip and become pores or porosity by being trapped by the solidifying solid front. In the case of A5083, many large bubbles are generated from the keyhole tip and move backwards widely. Most bubbles become pores or porosity, but some bubbles disappear from the molten pool surface at low welding speed. In the case of Type 304, bubbles move up backwards in the molten pool, but do not reach the upper part and result in pores or porosity in the lower part of the weld bead. Figures 5.2 and 5.3 show the X-ray transmission in situ observation photographs focusing movement of a W particle in the molten pool during YAG

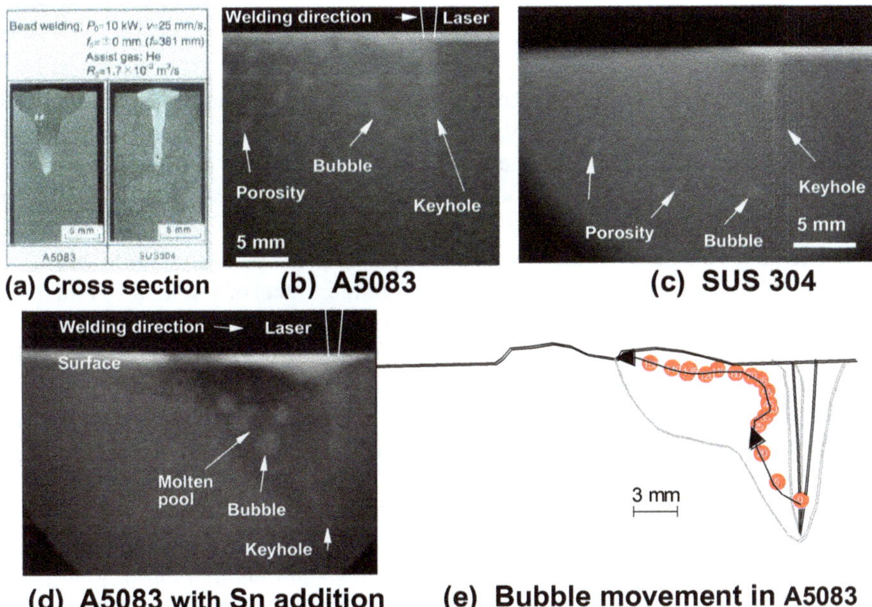

Fig. 5.1 Cross sections of CO_2 laser welds of A5083 aluminum alloy and Type 304 stainless steel, their X-ray transmission in situ observation photographs and example of bubble movement during laser welding of A5083 alloy

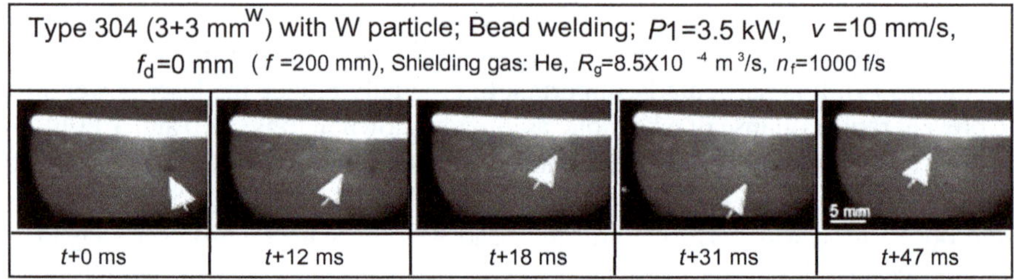

Fig. 5.2 X-ray transmission in situ observation photographs displaying bubbles formation and W particle movement in molten pool during YAG laser welding of Type 304 steel

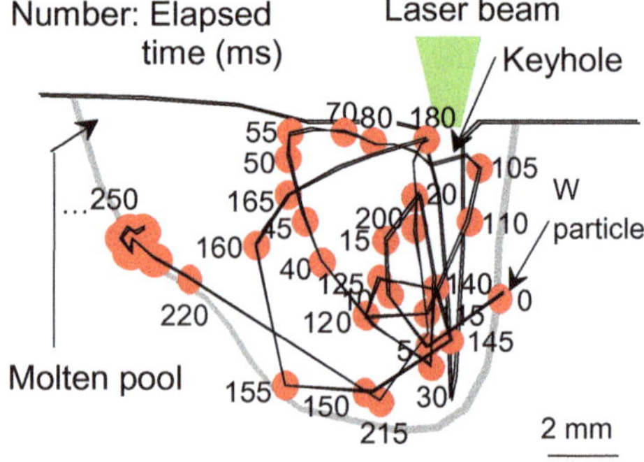

Fig. 5.3 Movement traces of W particle corresponding to melt flows in molten pool during YAG laser welding of Type 304 steel

laser welding, and movement traces of W particle corresponding to the melt flow in the molten pool, respectively [9]. W particles and bubbles move to circulate up backwards in the molten pool. The movement of W particle is similar to that of a bubble. It is consequently understood that bubbles move according to the melt flows in the molten pool.

Generally, full-penetration welding can produce porosity-free weld beads. In welding with CW CO_2 laser and YAG laser of stainless steel of less than 10 mm thickness, full-penetration weld beads without porosity could be produced by employing a full-penetration keyhole welding. However, porosity was present in full-penetration weld bead of HT780 steel plates of 10 mm thickness with focused fiber laser of 10 kW power, while sound full-penetration weld beads could be produced by using a modified focusing head of a slightly larger beam diameter and a slightly longer Rayleigh length, as already shown in Fig. 3.16 [10], and Fig. 5.4 shows a comparison of high-speed video photographs of molten pool motion and X-ray transmission in situ observation photographs during laser full-penetration welding between two optical heads [10]. Generally, as shown in the right photograph of Fig. 5.4, in full-penetration welding of a full-penetration keyhole, bubbles are hardly formed, resulting in no formation of porosity and the formation of a sound weld bead, since the solid zone or the melt zone does not exist below the keyhole tip. However, when the laser power density due to a small beam diameter is high, an extremely high

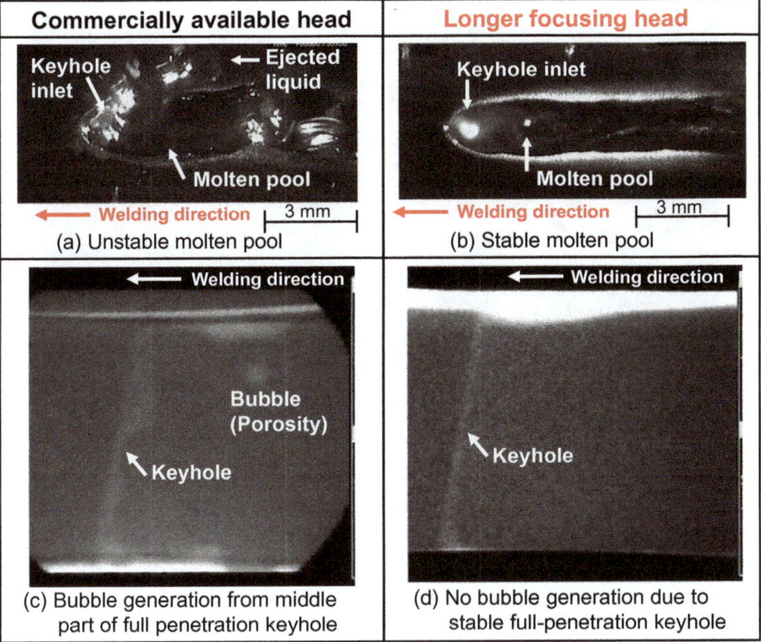

Fig. 5.4 Comparison of high-speed video observation results of molten pool and X-ray transmission in situ observation photographs during laser full-penetration welding between commercially available head and modified longer focusing head, showing unstable molten pool and keyhole leading to bubble formation from middle location of keyhole in case of commercial head, and stable molten pool and keyhole resulting in no bubble formation in case of modified head of wider spot beam size and longer Rayleigh length

power density beam can be shot on the keyhole wall below the plate surface and the keyhole becomes unstable due to the evaporation from the front wall, and consequently, the bubbles are generated in the rear molten pool from the middle depth part of the keyhole due to the severe evaporation at the front wall, resulting in the formation of porosity, as shown in the left photograph of Fig. 5.4 [10]. In high-power laser welding, it is preferable to use a moderately large beam diameter at the focal point for the prevention of porosity.

A molten pool, a keyhole and its behavior, bubbles generation and flows, and porosity formation were observed during welding with a single-mode fiber laser of 500 W power and about 25-μm-beam diameter at the speed of 1 m/min, the defocused distance of −1 mm and the inclination angle of 10° by using an ultrahigh intensity synchrotron radiation beam of SPring-8. An example of observation photograph of a molten pool and a keyhole during laser welding of A6061 aluminum alloy is shown in Fig. 5.5 [11, 12]. It is apparent that direct visualization of a molten pool and a keyhole is possible without any tracers, and bubbles are generated from the keyhole tip and flow up backward near the molten pool bottom. Most of bubbles diminish just before solidification. This result supports to interpret that bubbles are

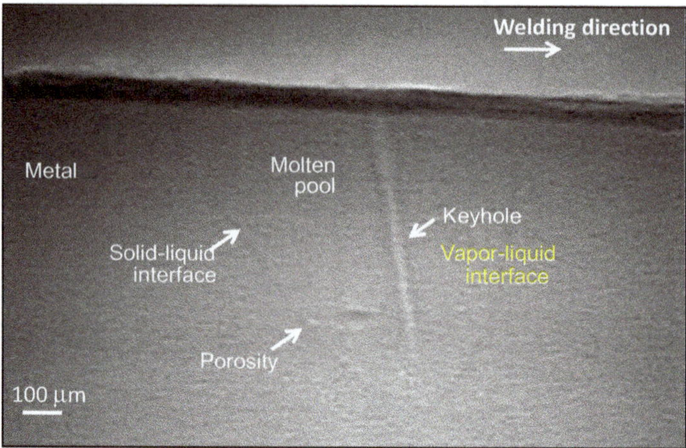

Fig. 5.5 In-situ observation photograph of molten pool during single-mode fiber laser welding of A6061 aluminum alloy observed through ultrahigh intensity beam of SPring-8, showing clear shapes of molten pool and keyhole

mainly composed of evaporated metallic vapors and do not include a shielding gas nor atmospheric gas components of nitrogen, oxygen, etc. Metallic vapors should change solid matters near the melting point, resulting in disappearance in the molten pool [11, 12]. On the other hand, if a shielding gas and atmospheric components are included in bubbles, they are retained as pores or porosity in the same formation mechansim as the porosity formed under some normal conditions.

According to the analytical results of porosity gases with the in-vacuum Q-mass method, the main gas inside the porosity is a shielding gas, and a small amount of hydrogen (H) and sometimes a small amount of nitrogen (N) are detected [2, 7–9]. It is generally considered that the bubbles may have a small amount of shielding gas and atmospheric gases due to the unstable keyhole motion in addition to the same evaporated vapors as a keyhole. Oxygen (O) is extremely active to form oxide particles or films with metallic vapors as the temperature falls. Consequently, in the porosity, a large amount of argon (Ar) or helium (He) shielding gas and a small amount of N are retained. H is also contained because it can diffuse in the molten pool and weld fusion zone and invade the porosity during and after laser welding [2, 7–9].

Bubbles and porosity formation situation during laser welding are schematically illustrated together with keyhole behavior and melt flows based upon the X-ray transmission observation and analytical results in Fig. 5.6 [2, 13]. In the case of the normal laser power density of usual beam diameters at the low welding speeds, as shown in Fig. 5.6a, an unstable keyhole is easily formed and leads to the formation of bubbles, since a laser beam is irradiated on the wall, which causes the melt downward flows to produce a bubble by closing the bottom keyhole. Many bubbles become pores or porosity, but some bubbles may disappear from the molten pool surface due to strong upper backward melt flows. A bubble is sometimes formed in the bottom

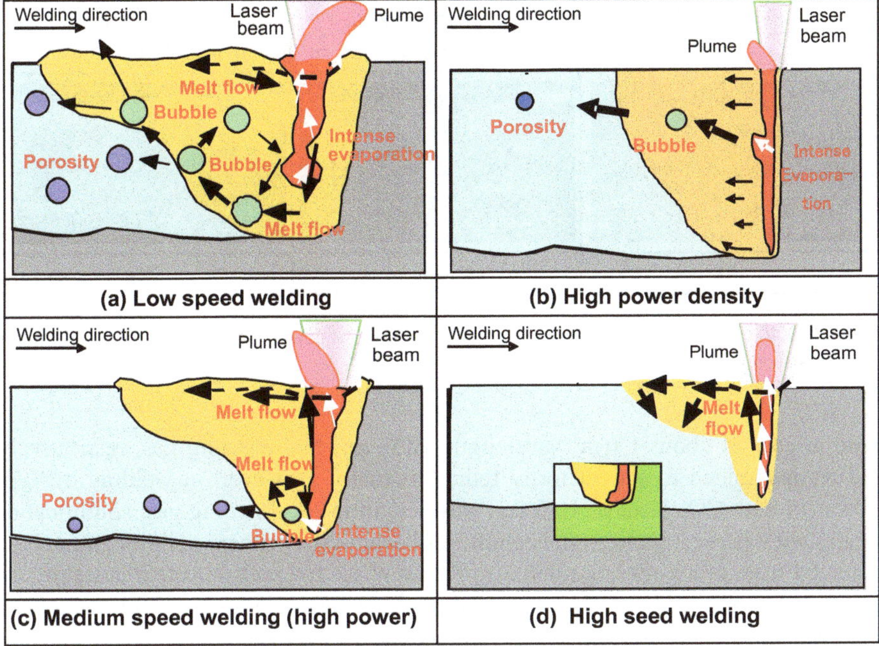

Fig. 5.6 Schematic illustration of bubbles and porosity formation situation together with keyhole behavior and melt flows during laser welding based upon X-ray transmission observation and analytical results

part of the molten pool when the laser power suddenly falls due to the interaction of the beam to a plume or the environment, etc. On the other hand, in the case of high power density of a small beam diameter, as shown in Fig. 5.6b, bubbles are generated near the upper or middle part of a keyhole due to severe evaporation and become pores or porosity. At the normal power density and the moderate welding speeds, as shown in Fig. 5.6c, bubbles are generated by strong evaporation from a keyhole tip wall and flow upper backwards. Most bubbles are trapped as porosity by the solidifying front. This formation mechanism is very often observed. At higher welding speeds, bubbles, resulting in pores or porosity, are smaller. Especially in steels and austenitic stainless steels at high welding speeds, as shown in Fig. 5.7d, bubbles formation can be suppressed probably due to plumes or evaporated vapors ejected upwards, and sound welds without porosity can be produced. It is revealed that bubbles are generated mainly from a keyhole tip under the normal conditions and partly from the middle depth of the keyhole when the power and its density are high. In some cases, the shielding gas affects the reduction in bubbles and porosity, as the X-ray transmission observation photographs during 3.5 kW YAG laser welding of Type 304 in Ar, He and N_2 shielding gas are shown in Fig. 5.7 [9, 13]. Bubbles and porosity are seen in Ar and He gas, while bubbles and porosity are hardly observed in N_2 gas. Accordingly, in normal welding of steels and austenitic stainless steels

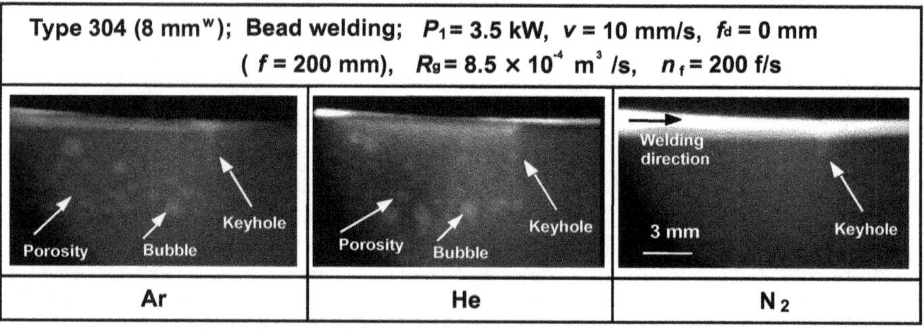

Fig. 5.7 X-ray transmission in situ observation photographs during 3.5 kW YAG laser welding of Type 304 steel in Ar, He, and N_2 shielding gas

with a laser of about 1 μm wavelength, CO_2 and N_2 shielding gas, respectively, are recommended to satisfactorily reduce porosity. However, in welding of Type 304 steel with CO_2 laser of 10.6 μm wavelength in N_2 shielding gas, solidification cracks were present under some conditions because the primary solidification phase changed from crack-resistant delta (δ) ferrite phase to crack-sensitive austenite (γ) phase as a consequence of the inclusion of a considerable amount of N (nitrogen) in the molten pool through a keyhole during welding. Some attentions should be paid because of easy formation of a plasma and incidentally N activation in CO_2 laser welding.

Oils, grease, and moistures on the joint surfaces or such matters of high vapor pressures trapped between joint surfaces may also induce spattering, underfilling, bubbles and porosity formation, and cold cracking. And thus, these matters should be wiped off and removed from the joint just before welding.

There are some alloys which are extremely sensitive to the formation of porosity. In laser welding of aluminum alloys, the removal of oxide films and hydrogen sources by cleaning butt-joint or lap-joint surfaces is needed for the reduction in porosity. In laser welding of steels, bubbles leading to porosity are likely to form due to the oxide films when the butt joints whose surfaces are prepared by laser cutting in O_2 gas are used. These bubbles may be generated by the evolution of CO gas by the rimming action (FeO+ C > Fe+ CO) between carbon (C) in the steel and O in the oxide films. In this case, deoxidation elements such as Al, Ti Mn, and Si may be effective to suppress or prevent this porosity. The steel plate surfaces are now prepared by laser cutting in N_2 gas or Ar gas owing to the suppression of oxidation and oxides formation.

Porosity or pores resulting from bubbles should be mainly formed from evaporation at the keyhole tip. Such porosity can be reduced or prevented by the following procedures:

- Keyhole full-penetration laser welding under the proper conditions of wide beam spot diameters,

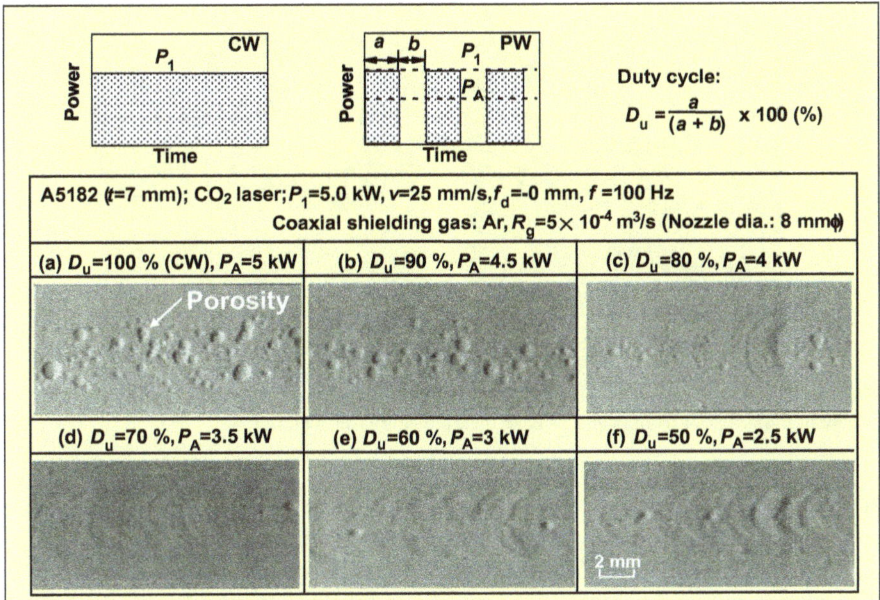

Fig. 5.8 Laser weld beads produced in A5182 alloy with CW and pulse-modulated CO_2 laser, showing effect of pulse duration and shaping on porosity reduction

- Welding with a shallow, stable keyhole-type welding or heat conduction-type welding, or in other words, welding with properly defocused conditions (for the formation of a conical, stable keyhole,
- Laser welding in low or high vacuum,
- Forward welding with an inclined laser beam,
- High-speed welding for steels,
- Pulse-modulated welding at low speeds, as shown in Fig. 5.8 [14, 15],
- Welding with twin laser beams or with a small-diameter laser beam with the addition of a ring mode beam,
- Selection of a proper shielding gas for each material, such as CO_2 gas for steels and N_2 gas for austenitic stainless steels.

In the case of cast alloys, sound welds can be produced in joining base plates with less macro-segregation, while poor welds with porosity or pores may be formed in welding base plates with macro-segregation including gases. In welding of die-cast or thixomolded alloys, small or large bubbles leading to pores or porosity are inevitably formed from the expansion of origins of very small bubbles of H, N, or CO_2 gas minimized by high pressure in the production process of the base materials. An example of porosity in laser weld and die-cast AZ91 magnesium alloy is shown in Fig. 5.9 [16, 17]. Such porosity is characterized by the growth of exceedingly small holes near the fusion boundary. It is therefore understood that the reduction in such

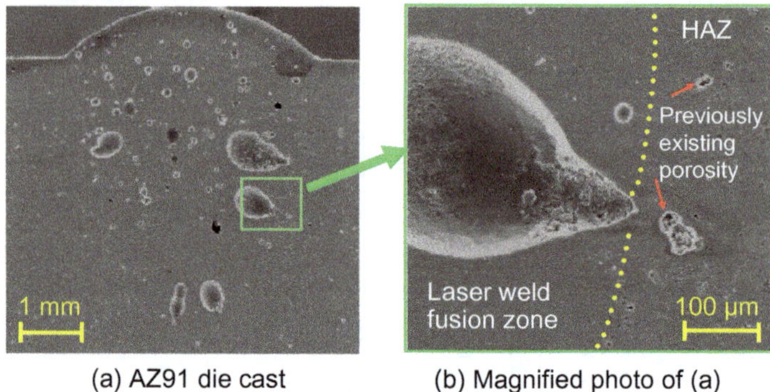

Fig. 5.9 Photograph of laser weld bead of die cast AZ91 magnesium alloy, showing expansion of bubble in molten pool from origin of base metal during laser welding

bubbles and porosity is extremely difficult in fusion welding. The use of insert sheet and two passes runs may be recommended for the reduction in porosity though.

In laser lap welding of Zn-coated steel sheets, a proper degree of gap between the sheets is required to stably produce a sound weld without porosity nor without underfilling due to severe spattering. The main phenomena encountered during laser lap welding of Zn-coated steel sheets are schematically represented in Fig. 5.10 [18]. In the case of no gap at low welding speeds, as shown in (a), bubbles are generated not only from the front fusion boundaries but also from the lateral fusion boundaries of the lap interface due to wide HAZs, and consequently, wormholes or large-sized porosity (blowholes) are often present. At normal high welding speeds, as shown in (b), spatters are likely to occur by evaporated Zn vapors, and consequently, rough surfaces or weld beads with large pores are formed. At proper slow welding speeds, as shown in (c), the formation of bubbles and porosity (wormholes) can be reduced due to narrower HAZs by selecting a proper pulse modulation. On the other hand, as shown in (d), in lap welding of sheets with a proper gap of about 0.1–0.2 mm, sound weld beads without porosity nor underfilling can be easily formed.

Concerning Zn-coated steel sheets, trials of laser butt-joint welding are less than ones of laser lap welding. It is noted that sound laser weld beads can be produced in butt-joint welding of Zn-coated steel sheets without gap. It means that the butt-joint welding is feasible for joining of Zn-coating steel sheets. It also signifies that Zn-coating on the plate surface hardly affects welding results of porosity and underfilling.

Laser brazing of Zn-coated steel sheets has been used in car industry. In laser brazing of Zn-coated steel sheets, sound brazed metals can be formed under the proper brazing conditions although porosity or pits may be produced at lower laser powers or at extremely high welding speeds. Examples of brazed metal with porosity and sound brazed metal are shown together with X-ray transmission observation photograph during YAG laser brazing in Fig. 5.11 [19]. In this brazing, the evaporation of Zn vapors from the Zn-coating layers on the flare joint beneath the brazing fusion zone is one main cause of bubbles and porosity formation. The formation mechanism of

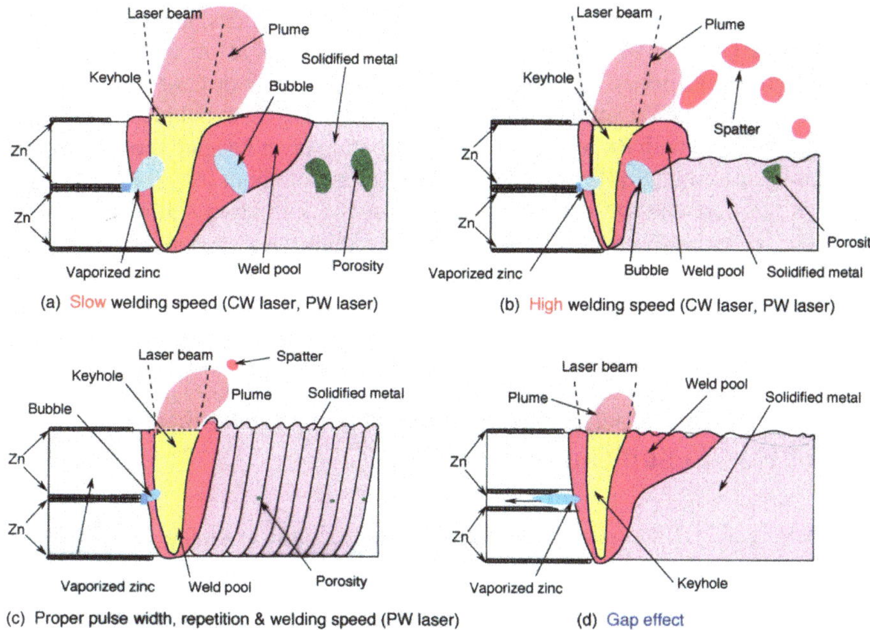

Fig. 5.10 Schematic representation of main phenomena during laser lap welding of Zn-coated steel sheets, showing bubble formation at low speeds (**a**), severe spattering at high speeds (**b**), and small porosity under proper conditions of pulsed laser (**c**) in lap sheets without gap, and no porosity and reduced spattering in lap sheets with proper gap (**d**)

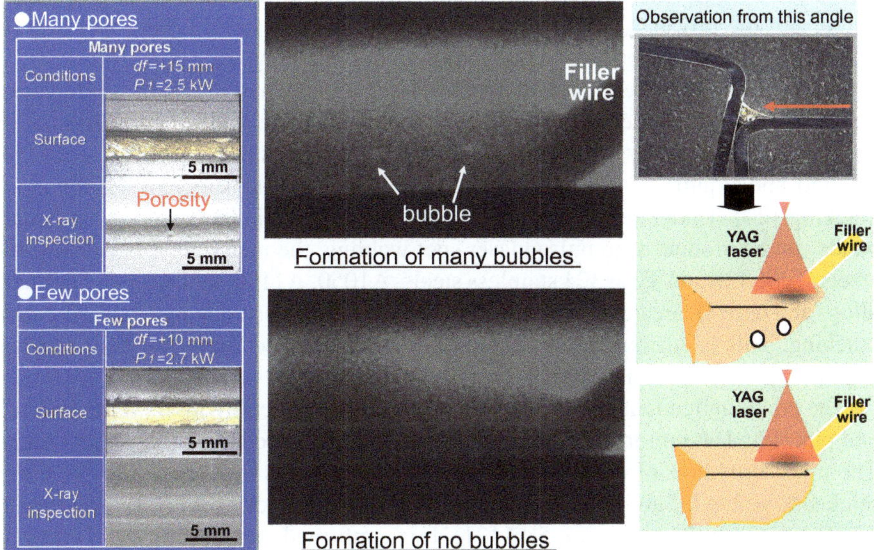

Fig. 5.11 Examples of brazed metal with porosity and sound brazed metal together with X-ray transmission in situ observation photographs during YAG laser brazing

porosity or pores in brazing is the same for any lasers. Besides, a pit is likely to be formed in the case of a small amount of brazing molten pool at low wire feeding rates or at high brazing speeds. A molten pool made of Cu-3%Si or Cu-8%Al filler wire was thought to solve Zn vapors, but Zn vapors seemed to form bubbles. The surface of pit was covered with Zn of vapors from Zn-coating layer beneath the molten pool.

In laser lap welding of Zn-coated steel, the setting of about 0.1 mm gap between sheets is the best and the easiest procedure for the formation sound weld beads. Moreover, in order to produce a sound brazing joint without porosity nor pits of Zn-coated steel sheets, it is important to form an ample molten pool by melting a filler wire sufficiently.

5.3 Formation Mechanism and Preventive Procedures of Hot (Solidification) Cracking

Hot cracking or high-temperature cracking related to liquid or melt is called "solidification cracking" in the weld fusion zone or weld metal and "liquation cracking" in the HAZ (heat-affected zone). Solidification cracking and liquation cracking may occur along grain boundaries in aluminum alloys, fully austenitic stainless steels, Ni-based superalloys, and so on. In particular, solidification cracking can occur easily in spot welding with a pulsed laser or in high-speed welding with a continuous wave (CW) laser. These causes are attributed to microsegregation and the resultant formation of low solidification temperature liquid films along the grain boundaries. Therefore, the selection of proper materials, the procedure to narrow the solidification temperature range, and the process to reduce external tensile load or strain during welding are important for the reduction and prevention in hot cracking.

In the case of spot welding with a pulsed laser, as shown in Fig. 5.12 solidification cracks are present under the normal welding conditions, and thus the irradiation conditions of tailing laser power so as to narrow the area of a mushy (coexistence of solid and liquid) zone should be preferably selected to suppress solidification cracking [20]. Under the controlled conditions of laser pulse shape, sound seam welds can be produced. In pulsed laser spot welding, the materials, such as SM 490 steels, Type 430 and Type 304 stainless steels, A1050, A1100, and A3003 aluminum alloys and pure Ti, are generally judged to be resistant to hot cracking or solidification cracking. This is attributed to narrow solidification temperature ranges of residual liquid droplets even if the microsegregation occurs.

The cracking tendency of A7N01 alloy was investigated by laser welding sheets and plates of 1–6 mm in thickness at various welding speeds from 25 to 150 mm/s. The microstructures of cross sections of laser weld beads in A7N01 alloy of 6 mm thickness at 25, 75, and 150 mm/s, and the results of crack absence or presence in weld metals are shown in Fig. 5.13 [21]. It is confirmed that solidification cracking is likely to occur at higher welding speeds in thicker plates except for 150 mm/s in 5 and 6 mm thick plates. No cracking at 150 mm/s is attributed to the formation of

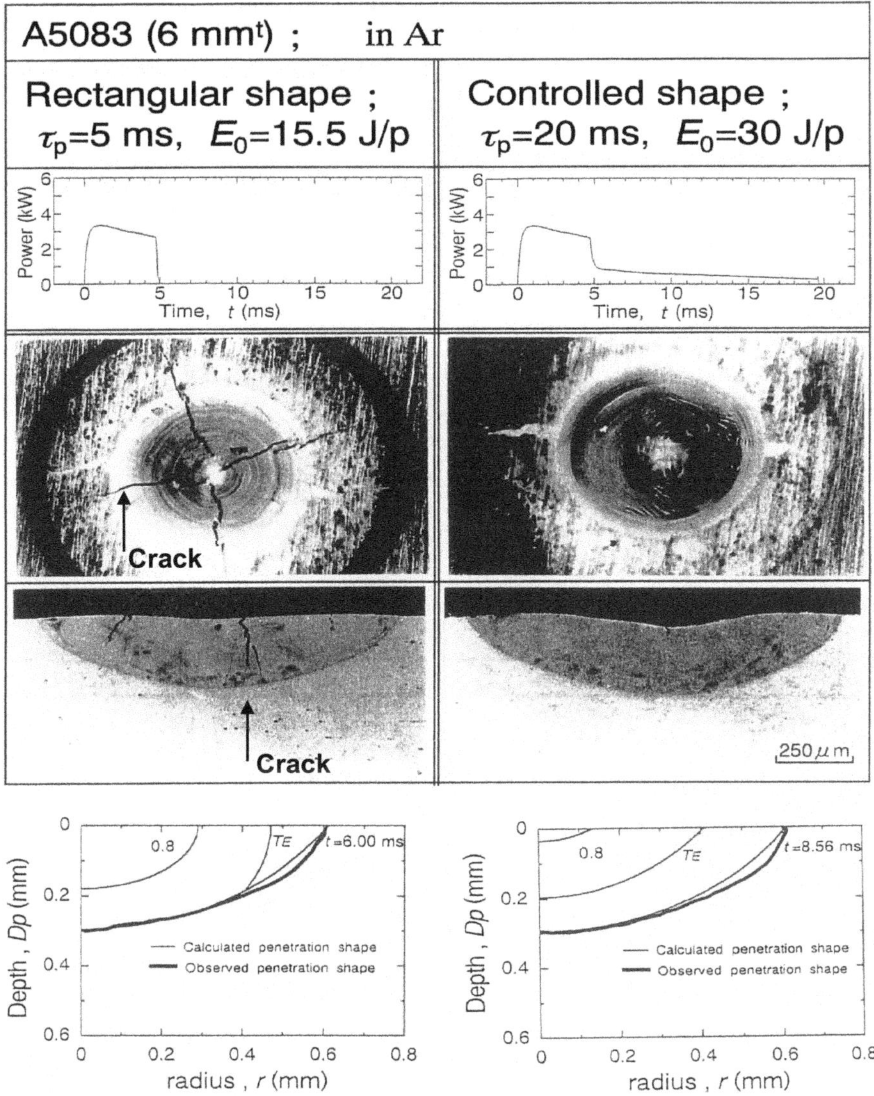

Fig. 5.12 Effect of tailing laser power on reduction in solidification cracking in spot weld fusion zone of A5083 alloy

a heat conduction-type of shallow, small weld fusion zones. It is interesting to know that the crack-free welds can be produced in the sheets of 1 and 2 mm thickness. Cracking tendency of laser weld fusion zones in the alloy is different depending upon the plate thickness and the welding speed. Nevertheless, it should be noted that when the sheets of 1–2 mm thickness are subjected to laser welding from the end of the sheet near the lateral side even in steels as well as aluminum alloys, cracking

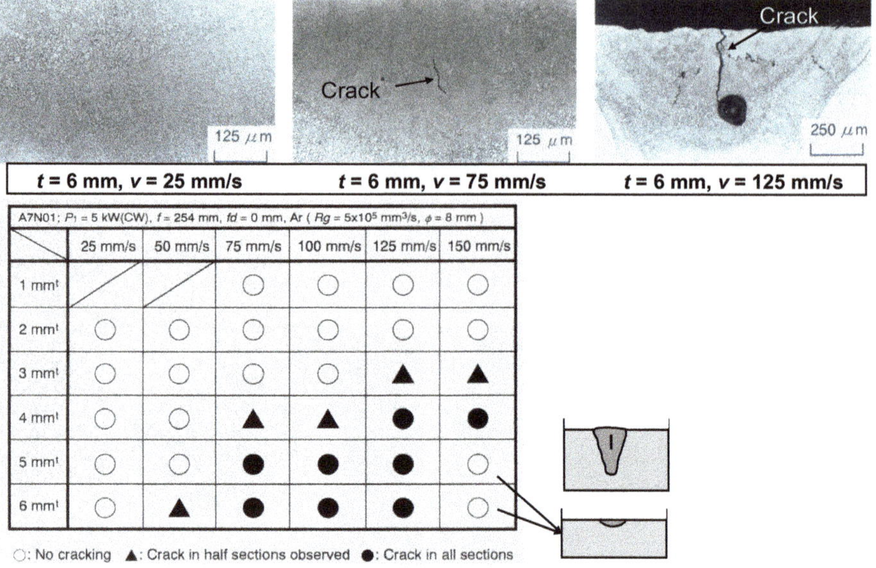

A7N01; P_t = 5 kW(CW), f = 254 mm, fd = 0 mm, Ar (Rg = 5x10^5 mm^3/s, ϕ = 8 mm)

	25 mm/s	50 mm/s	75 mm/s	100 mm/s	125 mm/s	150 mm/s
1 mm^t			○	○	○	○
2 mm^t	○	○	○	○	○	○
3 mm^t	○	○	○	○	▲	▲
4 mm^t	○	○	▲	▲	●	●
5 mm^t	○	○	●	●	●	○
6 mm^t	○	▲	●	●	●	○

○: No cracking ▲: Crack in half sections observed ●: Crack in all sections

Fig. 5.13 Cross-sectional microstructures of weld beads including no or some cracks made in 6-mm-thick A7N01 alloy plate with CO_2 laser at 25, 75, and 125 mm/s, and effects of plate thickness and welding speed on crack absence or presence in laser weld metals of A7N01 alloy

may take place due to the rotational deformation like fish bone-type or inverse fish bone-type cracking. The start location of laser welding is sometimes affective in joining thin sheets.

Steels are less sensitive to hot cracking or solidification cracking than aluminum alloys in CW laser welding. However, in fiber laser deep, partial-penetration weld fusion zones of thick steel plates, hot (solidification) cracking may occur. Examples of deep weld beads in HT 590 are shown in Fig. 5.14 [22]. In deeper weld beads made with fluxes or in low vacuum, such hot (solidification) cracking can be reduced or prevented by producing a deeper keyhole and by preventing retained long-range molten area due to the suppression of backward melt flows from the keyhole. Therefore, the causes may be attributed to the periodical formation of wide retained molten areas due to intermittent strong melt flows backwards near the bottom of the molten pool from the keyhole tip.

From the experimental results of laser welding, the relationship between hot cracking susceptibility, laser welding process, and the plate thickness is schematically summarized in Fig. 5.15 [2, 23]. Hot cracking susceptibility or solidification cracking sensitivity is the highest in spot welding with a pulsed laser, and solidification cracking occurs in most alloys. On the other hand, the cracking susceptibility is the lowest in CW laser welding of thin sheets (of about 1–3 mm thickness) under the proper welding conditions.

In order to reduce or prevent weld hot cracking, it is necessary to take the preventive measures from the viewpoints of both material or metallurgical and mechanical

Material		Surface	Cross section	Longitudinal section
Ceramics	Flux			
Use (11 mm gap)	No use		Hot crack	Hot crack area 20 mm
Use (11 mm gap)	Use			20 mm

Fig. 5.14 Examples of deeply penetrated weld beads in HT 590 steel made with 50 kW fiber laser, showing effect of flux (placed on plate surface) on increase in penetration depth and decrease in hot (solidification) cracking

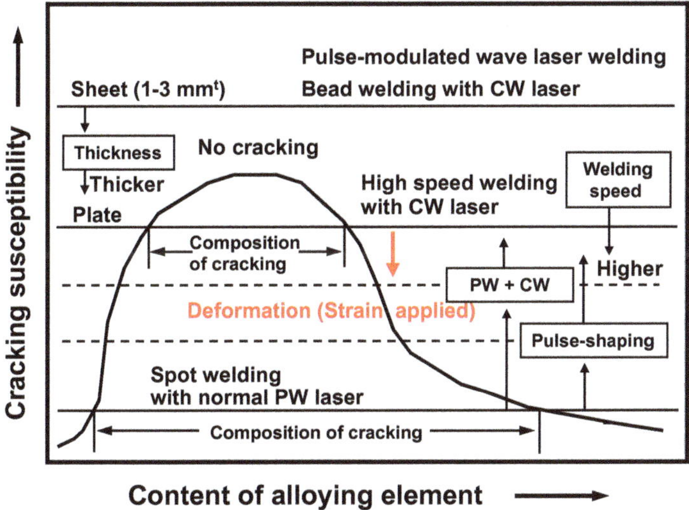

Fig. 5.15 Effects of laser welding process, welding speed, plate thickness, and alloying composition on weld solidification cracking sensitivity of certain alloy system

factors. Therefore, the following measures are considered in laser welding: (1) proper selection of the base metal or alloy, its thickness and weld fusion zone geometry, (2) proper control of molten pool compositions by using a filler wire, and (3) adaption of proper welding conditions for narrowing a mushy zone as well as suppression of rapid solidification and rapid tensile strain during solidification.

It has recently been reported in high tensile steels such as HT 980 that cold cracking may take place in the laser weld metals from the origin of solidification cracks in the crater. It is therefore important to reduce solidification cracking in any alloys and steels.

5.4 Formation Mechanism and Preventive Procedures of Spattering Leading to Underfilling

Spattering almost always occurs during laser welding of keyhole-type. In spot welding with pulsed YAG laser, spattering is likely to take place when the laser power increases very quickly, as already shown in Fig. 3.10 [24]. In bead welding, in particular, a large number of spatters are generated at high powers and high welding speeds, leading to underfilled weld beads, as already shown in Figs. 3.14, 3.15, 4.13–4.15, and 4.23. Moreover, in laser lap welding of Zn-coated steel sheets without gap, spattering occurs severely at high welding speeds, resulting in the formation of underfilled weld beads, as already shown in Fig. 5.10. To prevent spattering, control of the gap between sheets or removal of Zn-coated layer is needed.

A sound deep weld bead without underfilling could be produced in Type 304 steel at the power of 6 kW, the welding speed of 150 mm/s, and the defocused distance of −2 mm, as already shown in Fig. 3.15 Plume behavior, spattering, and melt flows in the molten pool were observed through high-speed video cameras and X-ray transmission observation system. The observation results of plumes and spattering from keyholes during laser welding at the defocused distances of +0 and −2 mm are shown in Fig. 5.16 [25]. And analytical results of melt flows upwards along with the keyhole wall during welding are compared in Fig. 5.17 [25]. It was understood under the conditions of a high power, a high speed, and a focal position that severe spatters occurred as a consequence of upper melt flows along the keyhole back wall due to strong plume (evaporated vapors) shear stream. At the defocused distance of −2 mm, upper flow rates of melts were slower than those at the focal point because of lower strong evaporation location, and consequently, large spatters could be greatly suppressed.

Figure 5.18 exhibits the surface appearances and cross sections of laser weld beads made at 6 kW and 9 m/min by changing beam inclination angles from -20° to +20° [25]. At the beam inclination angle of – side, severe spattering occurs, resulting in underfilled beads. On the other hand, at the angle of +20°, spattering is suppressed to form a sound weld bead. High-speed video and X-ray transmission observation results, analytical results of melt flows in the molten pool and the surface during welding and the number of large and small spatters at the beam incident angles of 0 ° (vertical) and +20° (forward) are compared in Fig. 5.19 [25]. A normal sound weld bead with reduced spatters is formed by suppressing the spout of large-sized melts in the upper directions only in the case of the laser beam of 20° forward inclination angle. It is apparent that such spattering is much liable to occur at the higher laser powers such as 10 kW. Nevertheless, the laser beam of 20° forward inclination angle

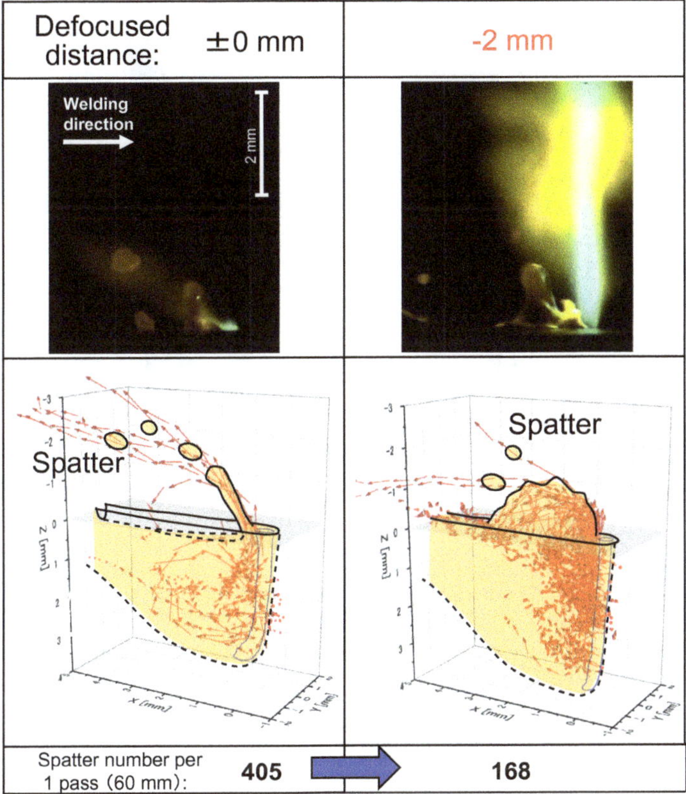

Fig. 5.16 Observation results of plumes and spattering from keyholes, moving speeds and direction of melt flows in molten pool and number of spatters during fiber laser welding of Type 304 steel at defocused distances of +0 and −2 mm

could reduce spattering and avoid the formation of underfilled beads at the high laser power of 10 kW.

The procedures for the reduction in spatters have recently been investigated by using special focusing optics such as ARM (adustable ring mode) laser of one small-diameter beam and its around ring mode beam. This laser machine, the optics, and the reduction results and mechanisms will be discussed in the following Sect. 6.6.

5.5 Formation Mechanism and Preventive Procedures of Humping and Undercutting

Humping is called when the periodical formation of humps of weld metal on the bead surface. Humping is likely to form in a narrow weld bead produced with an

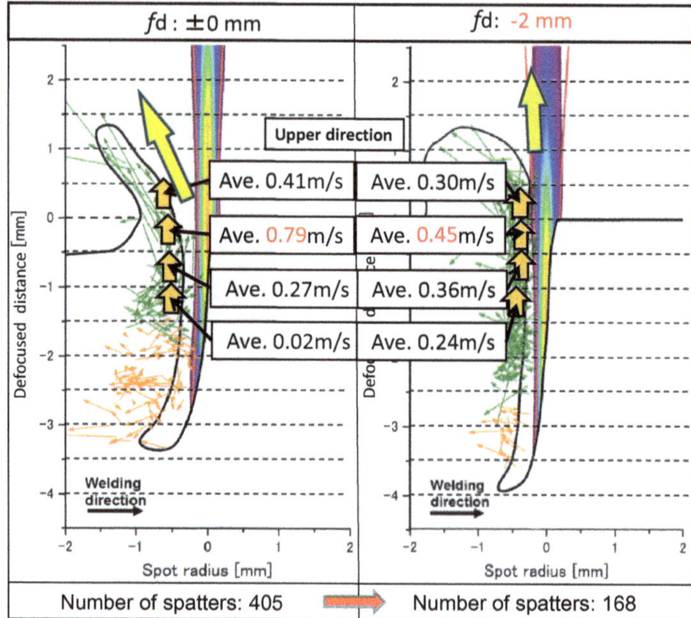

Fig. 5.17 Comparison of melt flows upwards along keyhole rear wall during fiber laser welding at defocused distances of 0 and −2 mm

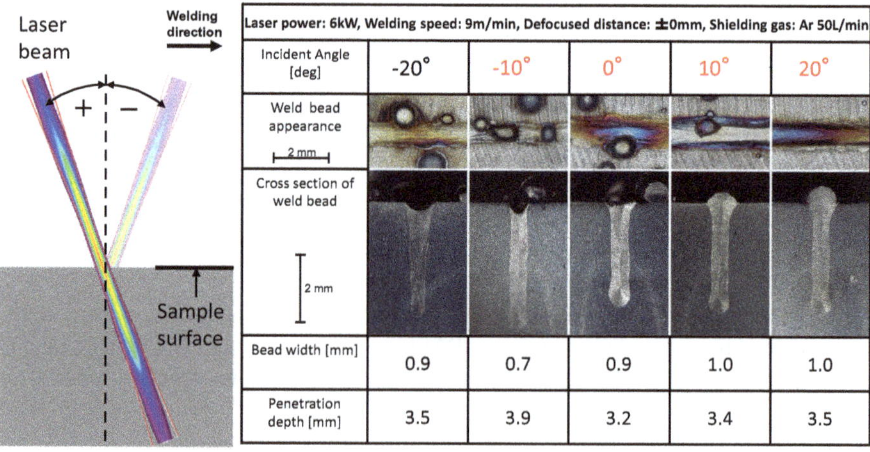

Fig. 5.18 Surface appearances and cross sections of weld beads made with fiber laser of 6 kW at 9 m/min and various beam inclination angles from −20° to +20°

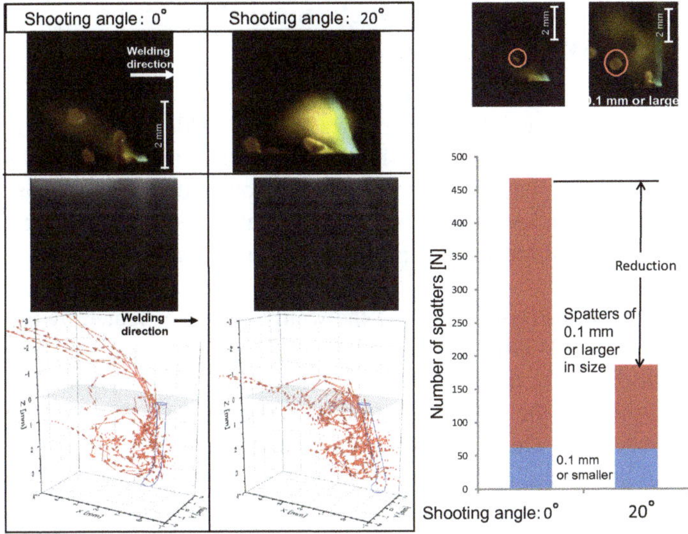

Fig. 5.19 High-speed video and X-ray transmission observation results, analytical results of melt flows in molten pool and surface during fiber laser welding, and comparison of number of large and small spatters at beam incident angles of 0° (vertical) and +20° (forward)

exceedingly small focused beam near the focal point during high-speed welding or during low vacuum welding. According to the observation results, humping was formed owing to the backward flow of a melt caused by plume (evaporated vapors) ejection and the high surface tension of the accumulated melt due to narrow molten pool width. Such humping can be suppressed by widening the molten pool width due to the use of a larger beam diameter or the change in the focal point below the plate surface, or by reducing the amount of melt ejected upward from keyhole inlet.

Undercut is a groove along the toe of a weld bead. Undercut is likely to occur in a wide bead made with a high-power laser, in the materials containing a high content of volatile elements, at high pressure of an assist (shielding) gas or under a large amount of flow of shielding gas from the front side of welding. It should be noted that the mechanical properties must be drastically reduced in aluminum alloys since the effective area for loading is reduced [15, 26]. Undercutting may occur easily in a full-penetration weld bead made at high arc current in laser–arc hybrid welding, because a high arc current induces shielding gas and plasma flows backward. In the case of fillet welding for a T-joint, an undercut is easily formed in the vertical plate. This is attributed to the effect of gravity.

Undercutting should be prevented by optimizing the welding conditions based upon its causes. In hybrid welding, an additional filler wire may be effective to prevent undercutting.

5.6 Hardness Profiles and Mechanical Properties of Laser-Welded Joints

Hardness profiles and mechanical properties of laser-welded joints are different from those of their base metals and alloys. Roughly speaking, the hardness of laser welds is higher than that of base metal in steels but lower than that of base alloys in aluminum alloys when laser melt-run (bead-on-plate) welding is performed without wires. Therefore, good mechanical properties of laser-welded joints are expected in steels.

The hardness of the base steels and the welds was investigated by using various industrial steels of 270–980 MPa class. Figure 5.20 shows the hardness profiles of the laser welds in comparison with those of arc and mash seam welds in 270–980 MPa steels [27]. Generally, the weld fusion zones and their neighboring HAZ correspond to the hardened zones. The hardened zones are narrowest due to the narrowest fusion zone width in laser welding. In the case of 270 and 440 steel, the laser weld beads are the narrowest in comparison with arc and mash seam ones, and the hardness of laser welds is the highest among welds. This is attributed to the formation of hard martensite and/or bainite phases due to the quenching of the weld metal and fast thermal history of the HAZ in laser welding. In the case of 590, 780, and 980 MPa steels, the base metals are hard and strong in the order of 590, 780, and 980 steel, and the highest hardness is obtained similarly in all welds, and the hardness levels become

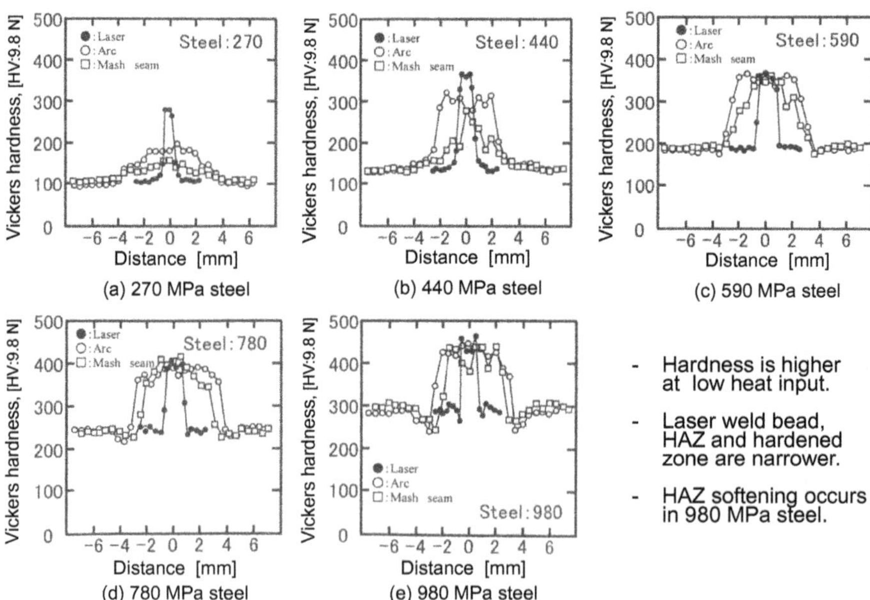

Fig. 5.20 Comparison of hardness profiles of laser, arc, and mash seam welds in various steels such as 270–980 MPa

high in the order of 590, 780, and 980 steel. In 980 MPa steel, the softening zones are present in both side HAZ at the lower temperatures. This softening zone is called HAZ softening. This softening zone should be formed in 780 MPa steel and higher tensile strength steels. In 780 MPa steel, slight HAZ softening is recognized due to higher heat inputs in arc and mash seam welding. HAZ softening is interpreted from the phenomenon that the hard martensite phase in the base metal should diminish to form tempering bainite phase including a small amount of soft ferrite phase along the lower temperature sides of the HAZ on the way of thermal history during welding. The softening levels are small in laser welds, and laser welding can narrow the zone and lessen the lowered levels since it can be performed at high speeds and low heat inputs. In the case of the joints possessing HAZ softening, fractures in the tensile test and the fatigue test should occur in the softening zone. Such softening is likely to occur in precipitation-hardening type of high tensile strength (HT) steels subjected to strong processing or deep forming. In 1500 MPa class HT steels, the weld beads and HAZ are soft, and HAZ softening is remarkable. Laser welding can reduce such HAZ softening at high speeds.

The hardness of various commercially available aluminum alloys and the laser welds was measured. Figure 5.21 shows the hardness profiles of the laser welds in different alloys [28]. In work-hardening alloys A3003-H114 and A5456-H116 represented as H (cold-working), as the base alloys are hardened, the HAZs are the softest, and the weld fusion zones become slightly harder than the HAZs. The softening of the weld fusion zones and the HAZ is understood by considering the

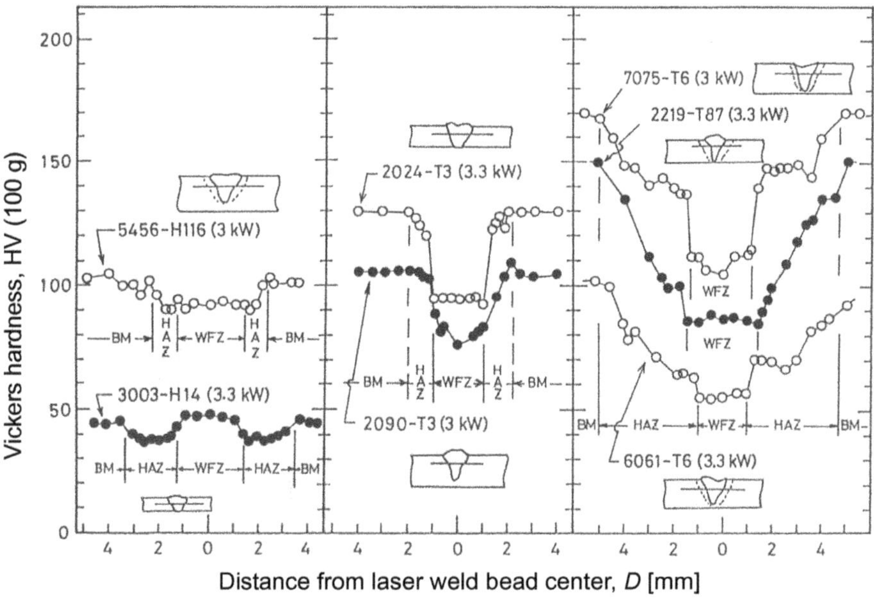

Fig. 5.21 Comparison of hardness profiles of CO_2 laser welds in different aluminum alloys

effect of microsegregation and strain–disappearance due to annealing, respectively. In precipitation-hardening alloys A2090-T3 and A2024-T3 represented as T3 (cold-working and then natural aging after solid-solution treatment), the weld fusion zones are the softest. In precipitation-hardening alloys A6061-T6 (artificial aging after solid-solution treatment), A7075-T6, and A2219-T87 (cold-working and then artificial aging after solid-solution treatment), the base alloys are the hardest but the weld fusion zones are the softest, and the hardness increases from the fusion boundary to the base metal in the HAZ. In the HAZ of T6 alloy, although the hardness increases gradually, the slightly softer zone exists on the way of gradual rise in hardness due to the overaging effect. In almost all aluminum alloys, the hardness becomes softer in the weld fusion zone or the HAZ than the base alloy.

Accordingly, concerning the mechanical properties of the laser-welded joints of aluminum alloys, the sound welds without welding defects are equivalent to the base alloys in the annealed conditions. In the other alloys, the mechanical properties of laser-welded joints are inferior to those of base alloys. The degraded levels are great in precipitation-hardening alloys.

In the case of Cu and Cu alloys, the plates are hardened by work-hardening, and thus, the hardness of the weld fusion zone and the HAZ is softer than that of the base metal just like H-type aluminum alloys.

References

1. Katayama S (2012) Ferrum (Bulletin of the Iron and Steel Institute of Japan) 17(1):18–29 (in Japanese)
2. Katayama S (2013) Defect formation mechanisms and preventive procedures in laser welding. In: Katayama S (ed) Handbook of laser welding technologies. Woodhead Publishing Limited, pp 332–373
3. Katayama S, Seto N, Mizutani M, Matsunawa A (2001) Proceedings of ICALEO 2001, LIA, Session C: Welding, 804 (CD), p 804
4. Katayama S, Kohsaka S, Mizutani M, Nishizawa K, Matsunawa A (1993) Proceedings of ICALEO '93, LIA, pp 487–497
5. Kim J-D (1997) Study of temporal and spatial high-resolution measurement, and formation mechanisms and preventive procedures of defects. Thesis for Doctor of Engineering (in Japanese)
6. Nakamura H (2015) Elucidation of fundamental phenomena in laser welding of titanium, and evolution to precision micro-joining. Thesis for Doctor of Engineering (in Japanese)
7. Katayama S, Seto N, Kim JD, Matsunawa A (1997) Proceedings of ICALEO '97, LIA, vol 83-Part2, Section G, pp 83–92
8. Katayama S, Seto N, Kim J-D, Matsunawa A (1998) Proceedings of ICALEO '98, LIA, USA, 84-Session C, pp 24–33
9. Katayama S, Seto N, Mizutani M, Matsunawa A (2000) Proceedings of ICALEO 2000, LIA, USA, 89-Session C, pp 16–25
10. Katayama S (2008) Report of the AMADA foundation, pp 240–245. http://www.amada-f.or.jp/r_report2/kkr/24/AF-2008218.pdf. (in Japanese)
11. Katayama S (2015) J Jpn Weld Soc JWS 84(8):582–590 (in Japanese)
12. Katayama S (2020) J Jpn Weld Soc JWS 89(1):5–15 (in Japanese)
13. Katayama S, Kawahito Y (2007) J High Temp Soc Jpn, HTSJ 33(3):118–127 (in Japanese)

14. Katayama S, Matsunawa A (1998) Proc CISFFEL 6–1:215–222
15. Katayama S (2006) J Light Metal Weld JLWA 44(8):333–343 (in Japanese)
16. Katayama S (2019) Very easy book of laser processing. The Nikkan Kogyo Shimbun, Ltd. (in Japanese)
17. Katayama's Labo.-2, JWRI, Osaka University
18. Katayama S, Wu Y, Matsunawa A (2001) Proceedings of ICALEO 2001, LIA, Florida, USA, P520, 1–9 (CD)
19. Kimura S, Takemura S, Mizutani M, Katayama S (2006) Proceedings of ICALEO 2006, LIA, USA, Paper #528, 346–354 (CD)
20. Katayama S, Mizutani M, Matsunawa A (1997) Sci Technol Weld Join 2(1):1–9
21. Katayama S, Ogawa K (2010) J Light Metal Weld JLWA 48(12):463–474 (in Japanese)
22. Sumimori D, Deguchi T, Nomura R, Watanabe K, Ashida Y, Katayama S (2018) Proceedings of the 89th Laser Materials Processing Conference, JLPS, 81–87 (in Japanese)
23. Katayama S (2009) Laser welding. J Jpn Weld Soc 78(2):124–138 (in Japanese)
24. Katayama S, Mizutani M, Ikeda H, Nishizawa K, Matsunawa A (1992) Proc ICALEO '92, LIA, 547–556
25. Kawahito Y, Nakada K, Uemura Y, Mizutani M, Nishimoto K, Kawakami H, Katayama S (2016) J Jpn Weld Soc JWS 34(4):239–248 (in Japanese)
26. Katayama S, Kojima K, Kuroda S, Matsunawa A (1999) J Jpn Weld Soc JWS 37(3):95–103 (in Japanese)
27. Yasuyama M, Uchihara M, Fukui K (2005) Proceedings of the 64th Laser Materials Processing Conference, JLPS, 52–59 (in Japanese)
28. Katayama S, Lundin CD (1991) J Light Metal Weld JLWA 29(8):349–360 (in Japanese)

Chapter 6
Characteristic Welding Processes

6.1 Tailored Blank Welding

Tailored blank welding was performed with laser welding, arc welding, or mash seam welding; but at present, it is predominantly carried out with laser butt-joint welding because it can produce the welded joints of good properties at high welding speeds.

Tailored blank welding with CO_2 laser started together with a computer numerical controlled (CNC) machine for manufacturing of large-scaled floor ban and sunroofs in the mid-1980s, then YAG laser welding was applied together with robot for tailored blank welding, and now tailored blank welding is applied for car doors and bodies by using disk laser, fiber laser or diode laser all over the world, as shown in Fig. 6.1 [1]. This tailored blank welding process consists of one process to manufacture one large blank plate by welding two to several sheets of different shapes, thicknesses, and strengths or chemical compositions, and then the other process to be subjected to the press-working of the blank plate. This tailored blank process is applied to weld general cold-rolled steel and high tensile strength steel, or different high tensile strength steels. The reductions of weight and cost are intended by arranging different sheets optimally and by reducing the number of parts. Consequently, the applications of tailored blank welding to car doors and bodies are diverse and increase.

Laser welds in steel sheets are generally hardened due to quenching; but in the case of HT steels of more than 780 MPa in tensile strength, HAZ softening is present along the HAZ near the base steel. This leads to the decrease in the fatigue strength of the joint. The formability of the welded joints is usually evaluated by the extrusion test. It is considered that the formability of the joints is low when the hardness of the welds is too high or the thickness of the weld bead is less than 0.8 thinner than that of the base sheets. Good or poor weld beads are judged by monitoring the weld bead surface geometry after welding or the gap between the sheets just in front of the molten pool during butt-joint welding, as shown in Fig. 6.2 [2]. When the gap is appreciably wide, an underfilled weld bead, which is susceptible to fracture during press-working, is formed, as shown in lower part of Fig. 6.2. The welded sheets made

S. Katayama, *Fundamentals and Details of Laser Welding*,
Topics in Mining, Metallurgy and Materials Engineering,
https://doi.org/10.1007/978-981-15-7933-2_6

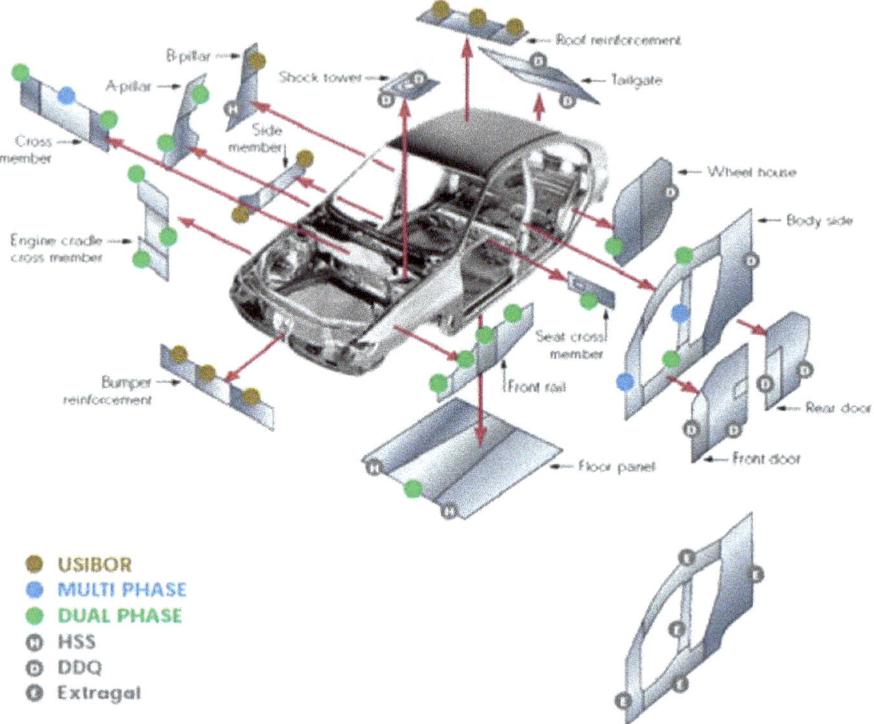

Fig. 6.1 Examples of tailored blank welding for car doors and bodies

under the condition of such a wide gap should be avoided before stamp forming. In the case of different thick sheets, laser irradiation location is important to produce proper weld beads, and thus a laser beam is shot on the thicker sheet side.

6.2 Remote (Beam Scanning) Laser Welding

Remote welding or scanner welding is most expected as a high-speed and high-production joining process. The technical word "remote (laser) welding" is used in this book. Focusing optics of long focal length is employed with or without scanners in remote welding, as shown in Fig. 6.3. At present, the system with scanners is mainly used [3–5]. Generally, a high-power and high-quality laser beam of several kW powers (4–6 kW in most cases) is introduced and is used together with lenses of long focal lengths. Consequently, a laser beam can be moved instantaneously from one welded point to another point, which can achieve high-speed and high-production welding. The remote welding can move a laser beam extremely quickly,

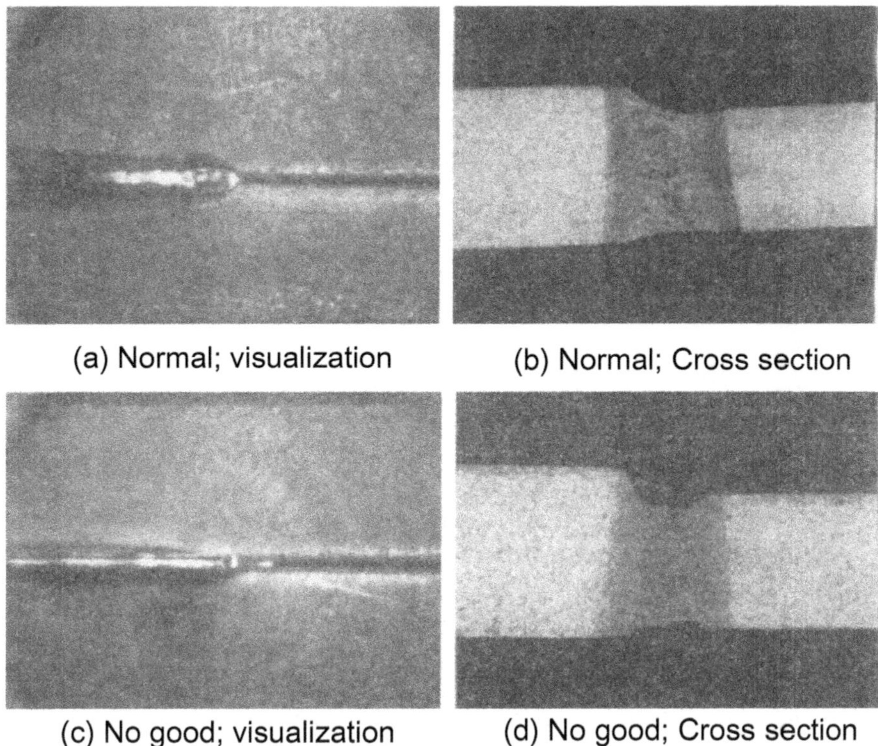

Fig. 6.2 Monitoring of gap between butt-joint sheets of different thicknesses in front of molten pool during fiber laser welding and cross sections of laser weld beads for quality evaluation, showing wider gap leading to underfilled bead judged as no good weld

cover a wide range of processing area, and produce many welds in a short time in a fixed posture.

In car industry, remote laser welding can be employed in the place of resistance spot welding and arc welding, and is applied to doors and car body parts. A slab-type CO_2 laser of high beam quality was initially used as a heat source of remote welding [6], and then LD-induced solid laser was used; but recently, fiber lasers or disk lasers are employed together with robots. It has been demonstrated that great reduction of operation time and labor number is feasible [7].

When remote welding is performed in air, the penetration depths of weld beads vary and sometimes become shallow on the way of welding, as already exhibited in Fig. 4.21 [8]. The cause is attributed to the formation of a wide high-temperature area of a small density, and accordingly, a small refractive index due to evaporated vapors and plume ejection from a keyhole inlet during remote laser welding, as already shown in Fig. 2.11a. In other words, the focal point of the laser beam is shifted below the plate surface, and/or the laser beam is refracted by the interaction of the laser beam to the high-temperature, small refractive index area above the molten pool, as

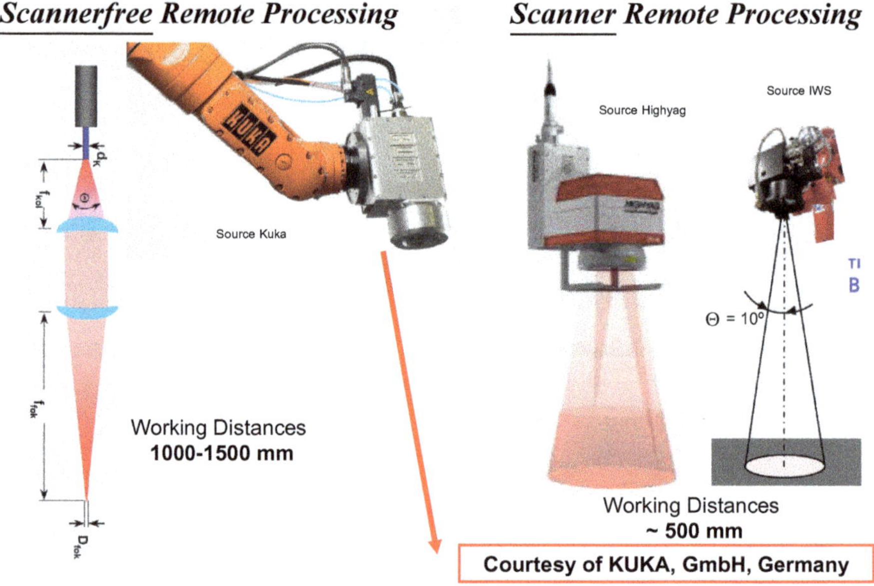

Fig. 6.3 Examples of remote laser welding with or without scanners

shown in Fig. 6.4 [8]. To produce a sound stable weld bead, the use of fan or blower and gas shielding is recommended with the objective of removing or reducing a low refractive index area of high temperatures during remote laser welding, as the steady formation of a full-penetration weld bead has already been shown in Fig. 4.22 [8].

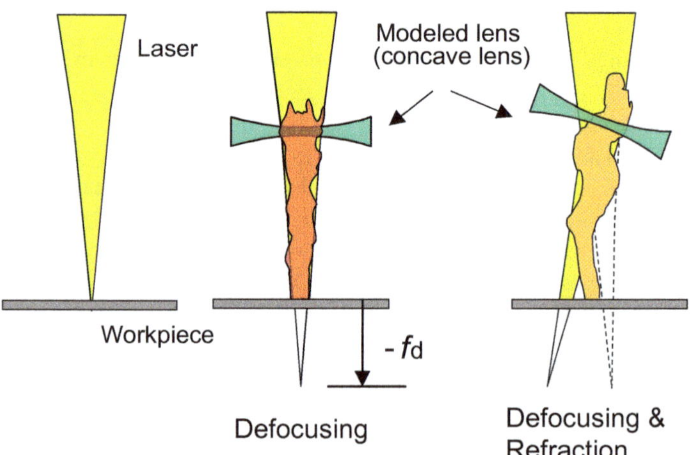

Fig. 6.4 Schematic illustration of plume behavior and its effects of defocusing and refraction due to the formation of high-temperature, small refractive index area above molten pool

6.3 Laser Brazing

Brazing, which is performed at the temperatures above 450 °C (723 K) by using a filler wire below the melting point of base material under the conditions to minimize the melting of base material, is one of the joining methods using wetting and capillary phenomenon. There are various names such as preplaced brazing, arc brazing, vacuum brazing, furnace brazing, Cu brazing, Ag brazing, etc., according to the heat sources, working procedures, and the kind of filler material.

Laser brazing using a laser beam as a heat source is chiefly employed to join Zn-coated steel sheets, where a laser beam produces a molten pool by melting Zn-coated steel sheet surface slightly and a filler wire gradually, as exhibited in Fig. 6.5 [9]. In the automotive industry, laser brazing can secure high car body accuracy because of small sheet deformation due to low heat inputs. Thus, laser brazing was first employed to integrate by joining the upper and lower parts of press-formed sheets of Zn-coated steel in a trunk lid in Europe and North America, and later in Japan, as shown in Fig. 6.6 [10, 11]. The issues of manufacturing limitations are solved by installing trim parts with laser brazing. Moreover, laser brazing is employed for car

(a) Laser brazing bead

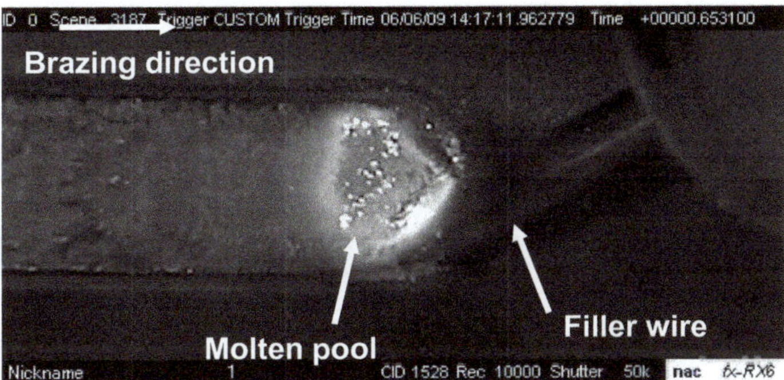

(b) Observation of molten pool and melting of filler wire during laser brazing

Fig. 6.5 Beautiful laser brazing bead and example of video observation during laser brazing of Zn-coated steel sheets

(a) Application examples of laser brazing to trunk lid, etc.

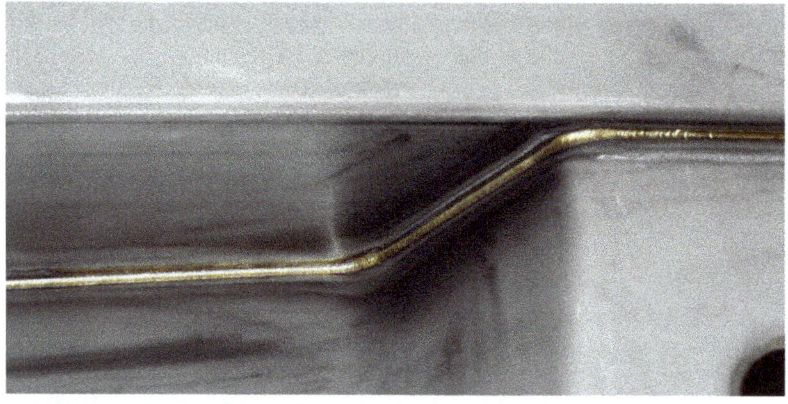

(b) Laser brazed bead applied to trunk lid

Fig. 6.6 Application examples of laser brazing to trunk lids of cars

body roofs, which brings the merits of the increased freedom of manufacturing as well as cost reduction due to the abolishment of sealing and molding, as exhibited in Fig. 6.7 [10]. Recently, laser brazing is used for joining of trunk lids and car bodies all over the world.

Lasers used at present are diode lasers, fiber lasers, etc. The filler wires for laser brazing are chiefly Cu-3%Sn, Cu-8%Al, and so on. Cu-based wires are used because they can contain a large amount of Zn. In Zn-coated steel, Zn is easily evaporated since the boiling point of Zn is lower than the melting point of steel (or Fe), resulting

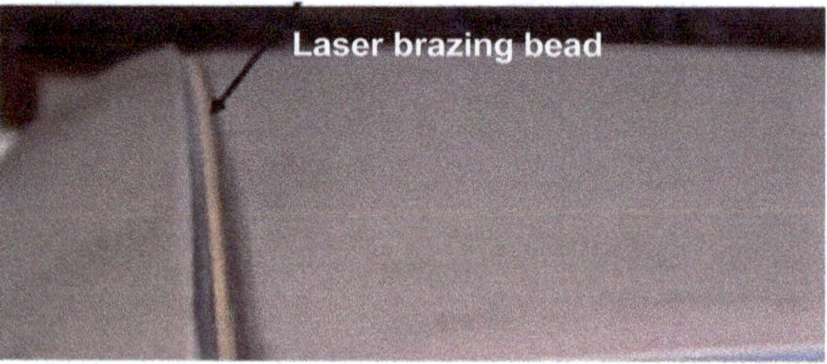

Fig. 6.7 Application example of laser brazing to car body roof

in the formation of porosity (pores) and pits in a brazed bead, as described in Sect. 5.2. The suppression of these defects is needed to increase the laser powers and the wire feeding rates.

Dissimilar metal joining was investigated by laser-brazed welding of aluminum alloy to steel, stainless steel, or Zn-coated steel. It is demonstrated that strong joints of dissimilar metals can be produced by laser-brazed welding under the proper joining conditions. In addition, the use of the flux to remove the oxides films covering the material surfaces is needed because the strong films such as Al_2O_3 and MgO covering aluminum alloys should hinder laser brazing.

6.4 Laser Soldering

Soldering, which is performed by using a filler wire below 450 °C (723 K) under the conditions to minimize the melting of base material, is one of the joining methods for supplying melts of a filler wire by using wetting and capillary phenomenon. Soldering joints are brittle and inferior in corrosion and have lower strengths than brazing ones. Conventionally, Pb-Sn alloys were used for soldering; but currently Sn-Ag alloys and so on are employed without Pb because Pb is harmful to human bodies. Pb-free alloys have higher melting points, resulting in the degradation in the degree of wetting. Nevertheless, laser soldering is hardly affected by such higher melting points of solders because a laser can easily melt the soldering wires.

Laser soldering is performed to join small parts by several methods of beam scanning in the electronics industry, as shown in Fig. 6.8 [12]. Laser soldering can be operated automatically, as exhibited in Fig. 6.9 [13]. The production of sound

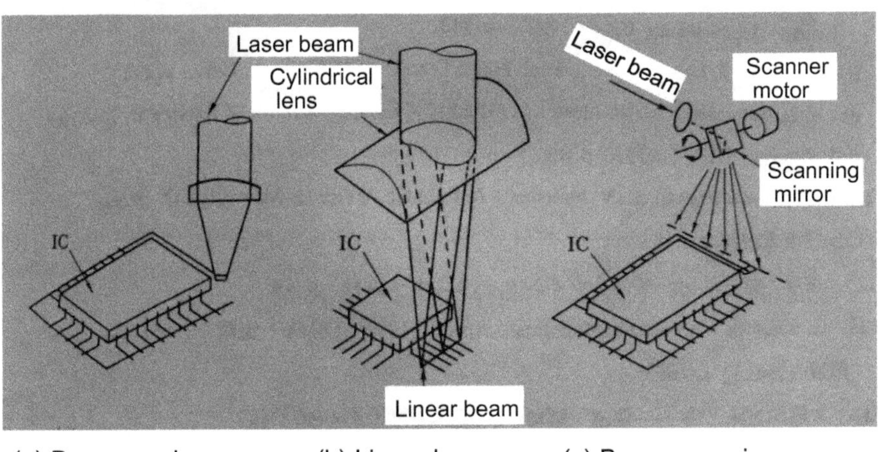

Fig. 6.8 Various beam scanning methods for laser soldering

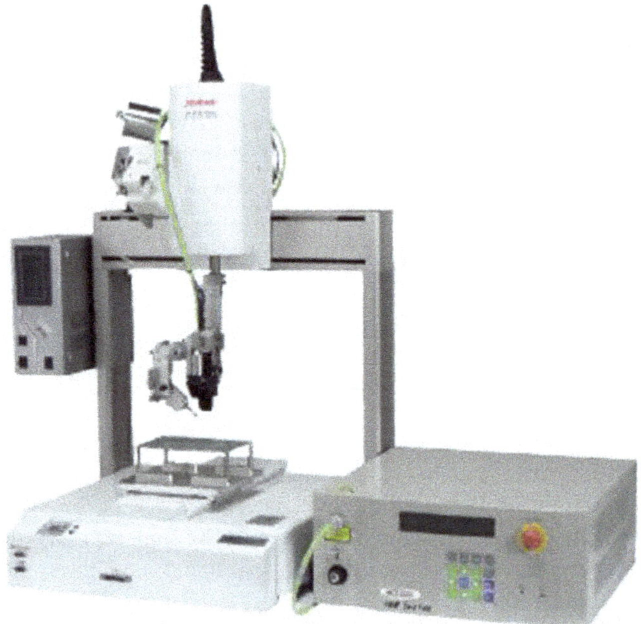

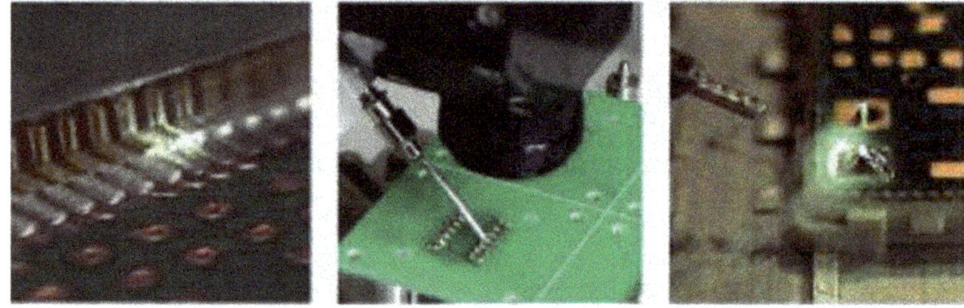

Fig. 6.9 Laser soldering apparatus and examples of laser soldering

good soldering joints should be carried out under the proper conditions of laser power density and irradiation (time) period. Diode lasers are mainly used to intend the reduction in the size and the improvement of the efficiency.

6.5 Laser Welding with Multi-Laser Beams

Laser welding with multi-beams has been investigated with the objectives of increasing the weld penetration depth and decreasing the welding defects. Two

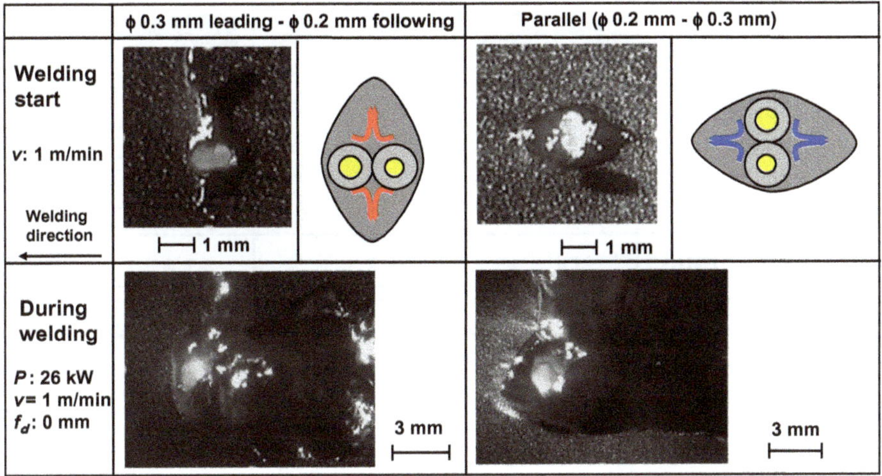

Fig. 6.10 Examples of high-speed video photographs after welding start and during welding of Type 304 stainless steel with two disk laser beams in inline and parallel direction, showing melts ejected from middle position between two keyholes in vertical direction against two beams

or three laser beams of the same wavelengths [14], two laser beams of different wavelengths [15], etc., are used as multi-beams.

Laser welding was performed by changing the power ratio of two beams using inline beam arrangement and cross-beam arrangement. In the case of the formation of two keyholes, the melts are ejected from the middle position between two keyholes in the vertical direction against two beams, as shown in Fig. 6.10 [14]. The penetration of the laser weld beads increases with an increase in the total power applied. In the case of inline beam arrangement, the welds of deeper penetration can be produced by arranging the following beam at higher power or higher-power density than vice versa arrangement. Figure 6.11 shows the effect of the power ratio of two laser beams (of 16 kW total power) on the weld penetration in Type 304 stainless steel at the welding speed of 1 and 3 m/min [14]. The penetration of a weld bead made with two beams is shallower than that with one beam at the same total power since the power density of two beams is lowered than that of one beam. In these experiment of deep weld beads, the effect of two beams on the reduction in welding defects is hardly remarkable.

The effect of two overlapped beams of different wavelengths was investigated. Figure 6.12 shows the cross-sectional photographs of seam welds produced with respective single beams and combined beams of pulsed YAG laser of 1.06 μm wavelength and Q-switched YAG laser of 0.53 μm wavelength [15]. The prominent effect of two overlapping beams on the increase in weld penetration is demonstrated. It is apparent that Q-switched YAG laser can conduce to the increase in the penetration of pulsed YAG laser weld beads, because YAG laser energy should be preferably absorbed into a deep keyhole made with Q-switched YAG laser.

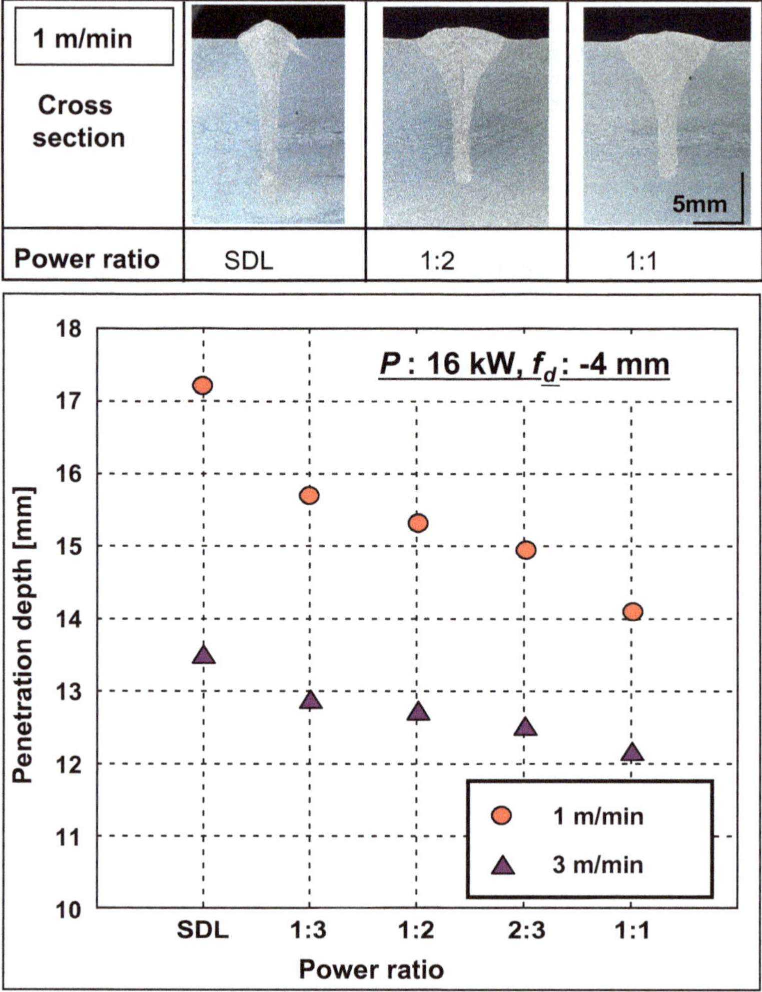

Fig. 6.11 Cross-sectional photographs of weld beads made with two disk laser beams of 16 kW total power, and effect of power ratio of two laser beams on weld penetration in Type 304 stainless steel produced at 1 and 3 m/min

Recently, the combined effect of a CW blue diode laser of wider-diameter beam and low power density and a normal CW diode or CW fiber laser of smaller-diameter beam and higher-power density was also investigated. Some results are already described in detail in Sect. 3.7. Figures 6.13 and 6.14 exhibit newly developed combined laser apparatus and the weld beads obtained with respective laser sources [16]. It is apparent that a stable deep sound weld can be produced with two beams of blue laser and fiber laser. Easier melting of metal due to the blue laser of

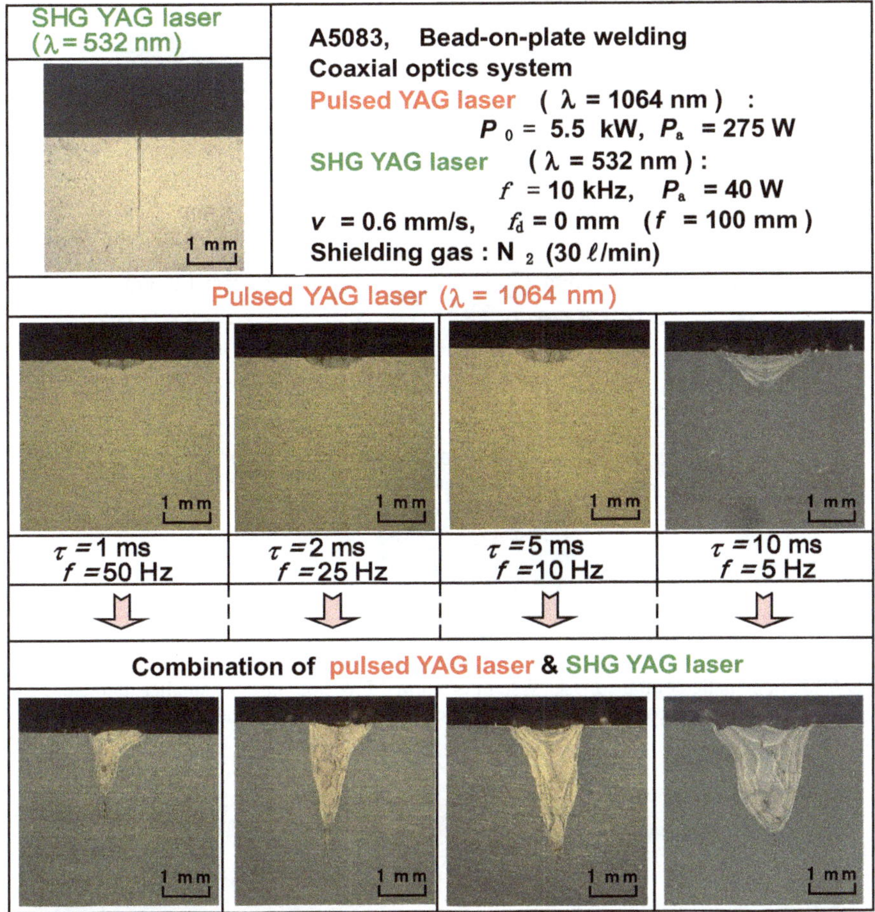

Fig. 6.12 Cross-sectional photographs of seam welds made with single beams and combined beams of pulsed YAG laser of 1.06 μm wavelength and Q -switched YAG laser of 0.53 μm wavelength, showing overlapping effect on the penetration of weld beads

shorter wavelength helps to the absorption of the laser of about 1 μm wavelength, resulting in the formation of a deep keyhole.

6.6 Laser Welding with Beam Mode-Modified Lasers

Recently, the procedures for the reduction in spattering are intensively investigated by utilizing double fibers or special focusing optics which can adjust power density profiles properly all over the world. The beam modes of single, ring, and double fibers are compared in Fig. 6.15 [17]. Phenomenon during welding with a single

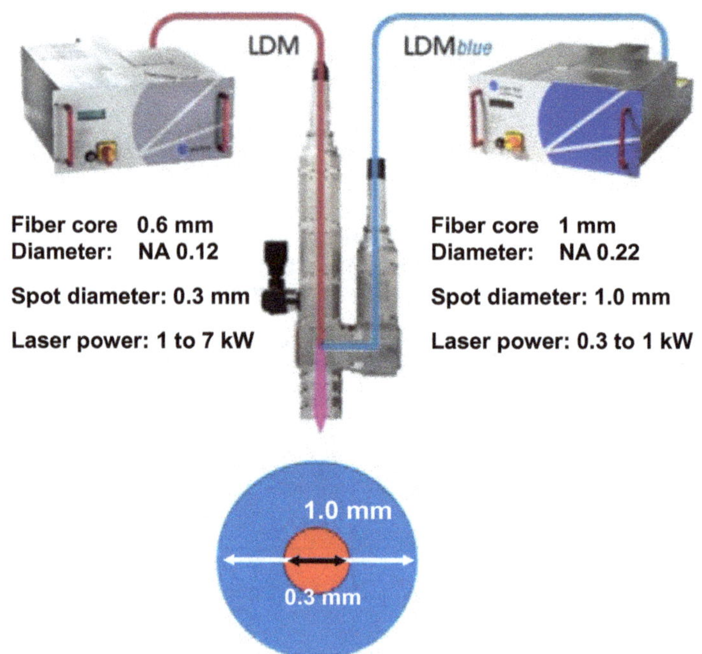

Fig. 6.13 Schematic combined laser system of normal CW fiber laser of smaller-diameter beam, high power, and high intensity, and CW blue diode laser of wider-diameter beam and low power density

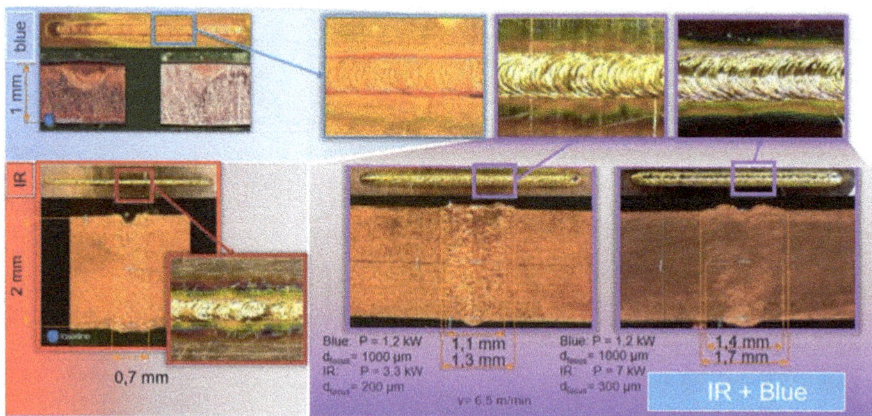

Fig. 6.14 Comparison of surfaces and cross sections of weld beads made in Cu plates with CW blue diode laser, normal CW fiber (IR) laser and combined laser of blue laser and IR laser, suggesting performance of stable welding with combined laser beams

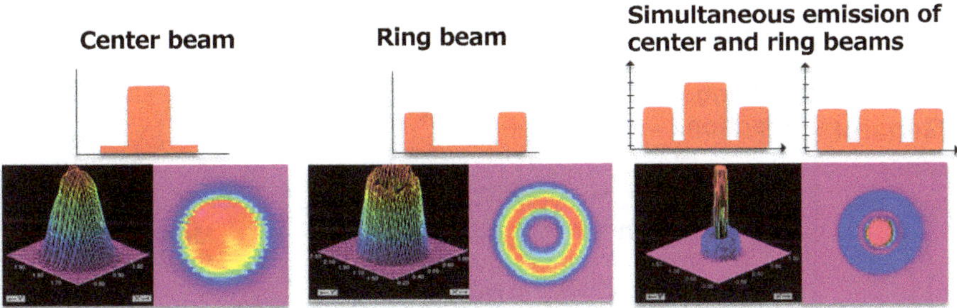

Fig. 6.15 Schematic and measured examples of beam modes (laser power density profiles) of single, ring, and double fibers for ARM (adjustable ring mode) laser

laser and single + ring mode laser (adjustable ring mode laser; ARM laser) is observed by normal and high-speed video cameras. The general and high-speed video photographs showing a molten pool, a keyhole inlet, and spatters during welding of A6061 aluminum alloy with a single laser and a single + ring laser are exhibited in Fig. 6.16 [17]. A molten pool is unstable during welding with a single laser beam and spatters occur violently, while in the case of a single + ring mode (ARM) laser the molten pool is stable and there are few spatters. Welding phenomena during

Fig. 6.16 Normal and high-speed video photographs showing molten pool, keyhole inlet and spatters during welding of A 6061 aluminum alloy with general laser and ARM (single + ring) laser

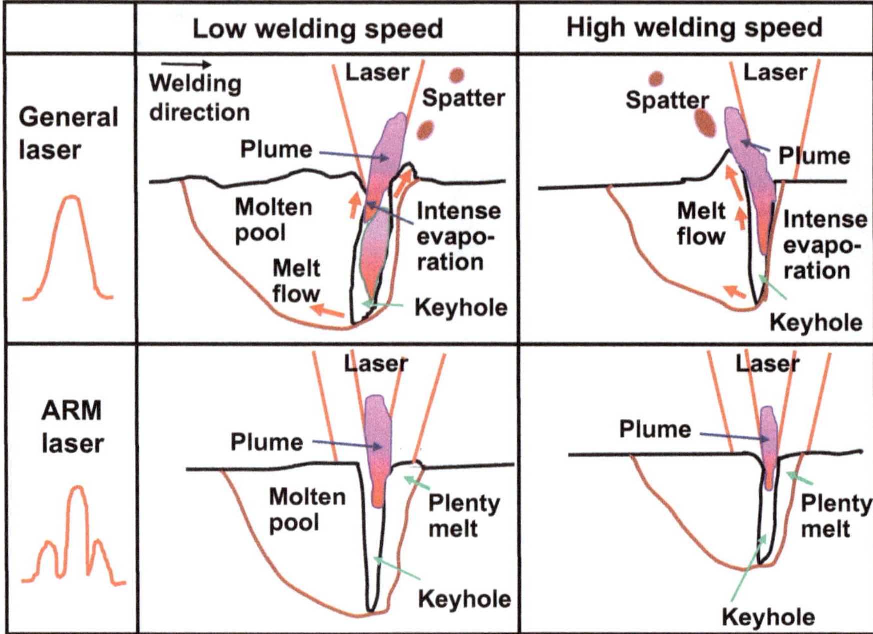

Fig. 6.17 Schematic illustration of welding phenomena during laser welding with general and single + ring mode (ARM) laser at low and high welding speeds, showing spatter suppression due to plenty of melts present in front of keyhole and formation of good weld bead surface due to suppression of backward melt flows

laser welding with a single and a single + ring mode (ARM) laser at low and high welding speeds are schematically illustrated in Fig. 6.17. The addition of a ring mode laser enhances melting of the front wall of a keyhole, resulting in the great reduction in spattering, and moreover, the rear part of the ring mode can suppress the melt flows backward from the keyhole inlet, leading to the good surface appearances. The beneficial effect of a single + ring mode (ARM) laser on the weldability, especially the suppression of spattering, is confirmed in welding of aluminum alloys, steels, Zn-coated steels, stainless steels, Cu, and so on. It has been reported that the other modified lasers can also reduce spattering and produce welds of good surface appearances. It is important to select the optimum laser beam mode in any cases.

6.7 Laser–Arc Hybrid Welding

Laser–arc hybrid welding has been noted as promising joining processes since it can compensate for the drawbacks or weaknesses in laser welding and arc welding by making good use of both features. Hybrid welding with CO_2, YAG, diode, disk or fiber laser and TIG, MIG, MAG, plasma or CO_2 gas arc heat sources has been investigated, and some combinations have been used in industries. Conventionally hybrid welding with CO_2 laser and TIG arc of steels started with the objective of increasing the weld penetration made by high-cost laser welding with a cheap arc heat source, and it was followed by hybrid welding with YAG laser and TIG arc, YAG laser and MIG arc and CO_2 laser and MAG arc of stainless steels or steels, aluminum alloys, and steels, respectively. Recently, hybrid welding with disk or fiber laser and MIG of stainless steels or aluminum alloys, or with disk or fiber laser and MAG or CO_2 gas arc of steels has been performed. These hybrid welding processes offer many advantages such as (1) deeper penetration, (2) higher welding speeds, (3) wider gap tolerance, (4) better weld bead surface appearances and (5) reduced welding defects leading to a smaller amount of porosity and a decrease in hardness in addition to complements of the drawback of both individual processes. Generally, two heat sources of a laser and an arc are used separately although the laser–arc coaxial torches were developed. Thus, the arrangement of laser and arc heat sources is important. In this book, when a YAG laser optics or TIG arc torch is a leading heat source, YAG-TIG or TIG-YAG hybrid welding is called depending upon the heat source arrangement. Examples of a laser weld and hybrid welds with TIG-YAG laser and MIG-YAG laser are shown in Fig. 6.18 [10]. It is understood that the use of a wire in MIG is advantageous for the prevention of underfilling leading to gap tolerance.

Laser-TIG hybrid welding situations are schematically shown in Fig. 6.19 [18, 19]. A keyhole is generally formed in a molten pool with a laser beam of high power density, and simultaneously plumes (light emission) of evaporated vapors or metallic atoms emitted from a keyhole inlet and from an arc-concentrated part near the keyhole inlet and spatters are formed over the keyhole. Besides, clusters and ultrafine particles or fumes are also formed from evaporated vapors in the space. In the molten pool, the melt flows downwards along a keyhole wall (in slightly poor

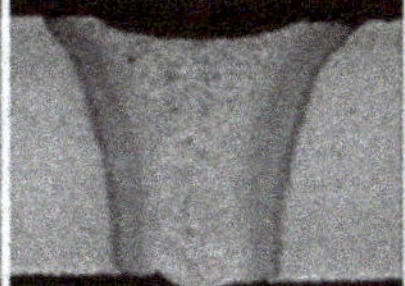
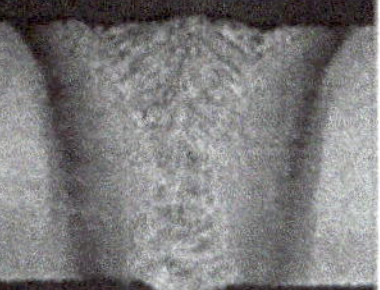

Fig. 6.18 Examples of cross sections of YAG laser weld bead and hybrid weld beads with TIG-YAG laser and MIG-YAG laser in steel butt-joint plates of 0.2 mm gap

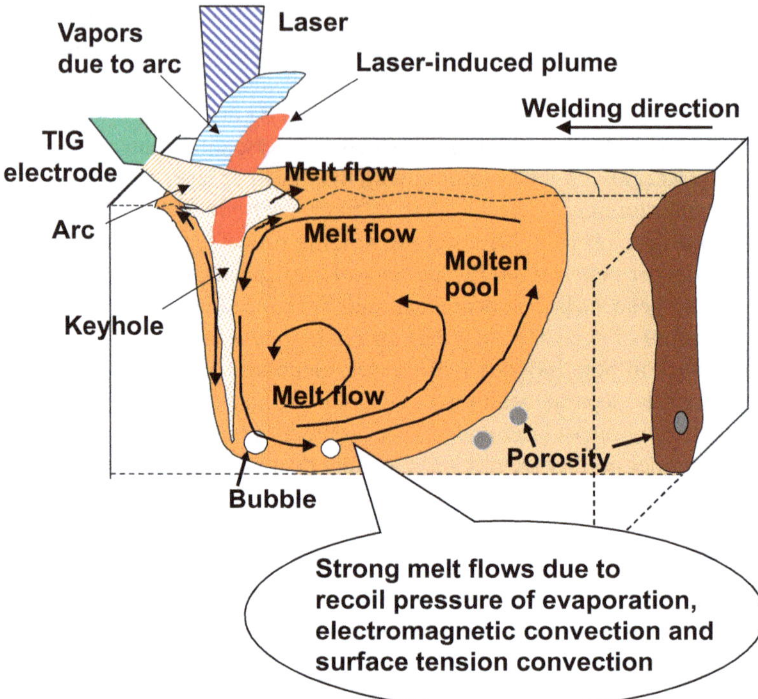

Fig. 6.19 Schematic illustration of laser-TIG hybrid welding phenomena

shielding conditions) and then from the keyhole tip near the molten pool bottom toward the rear part. These melt flows are attributed to three actions: (1) the recoil pressure on the keyhole wall due to the laser irradiation, (2) the stronger surface tension at higher temperatures in the case of the slightly poor shielding conditions, and (3) the electromagnetic convection due to an arc concentrated near the keyhole inlet. Consequently, a large molten pool is formed. Under some conditions, bubbles are generated in the molten pool from the keyhole tip and retain as pores or porosity in the weld fusion zone.

Under the conditions of high arc currents, the upper part of the weld bead is wide, as the video observation result during YAG laser–TIG arc hybrid welding of Type 304 stainless steel at the arc current of 200 A and schematic representation of hybrid welding situation are shown in Fig. 6.20 [18, 19]. At 100 A, a deep weld bead is formed by the formation of a deep keyhole and the downward melt flows due to recoil pressure, electromagnetic convection, and surface tension convection as already shown in Fig. 6.19, while at 200 A the molten surface is considerably concave, and the melt flows lateral back near the molten pool surface. The formation of wide-surface weld bead is attributed to the high arc pressure and the lateral back shear stream of a shielding gas. Besides, at 200 A, bubbles generation leading to porosity formation is reduced probably because a strong gas shear stream and the

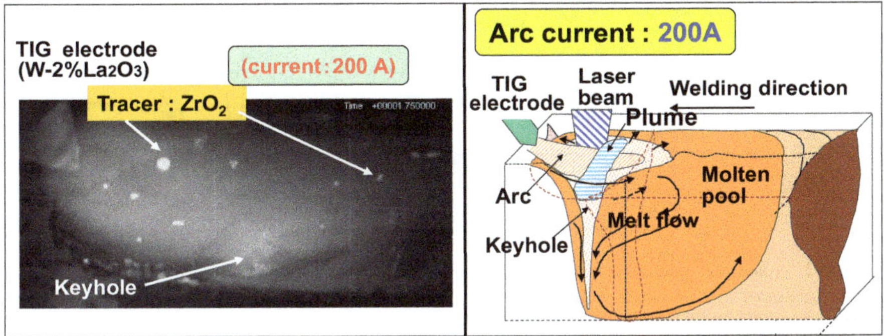

Fig. 6.20 High-speed video observation result of molten pool and flows of ZrO_2 particles, and schematic phenomena of molten pool, plume, and melt flows during TIG arc–YAG laser hybrid welding of Type 304 stainless steel at arc current of 200 A

concave surface acts as the prevention of shielding gas invasion into the keyhole and the suppression of bubbles generation under the defocused conditions leading to a shallow keyhole

The distance between a laser spot and an arc electrode target in each arrangement affects the weld penetration and geometry. In one experiment, for example, a relatively long distance of about 5 mm and a short distance of 1–2 mm are the deepest in TIG-YAG and YAG-TIG hybrid welding, respectively. It is the most effective when the laser beam is shot on the arc-concentrated potion. In the case of TIG-YAG hybrid welding, the deeper penetration can be obtained if the laser can be irradiated on the molten pool made with TIG arc. However, in the case of YAG-TIG hybrid welding, the weld is the deepest when TIG target is just behind the laser spot but becomes shallow when the TIG target is several mm behind the keyhole. In the latter case, the electromagnetic convection helps to produce a shallow weld by hindering melt flows backwards to form the keyhole tip.

In welding of aluminum alloys, hybrid welding with a laser and MIG arc is preferably used to produce a stable weld bead. MIG arc is performed in pulsed mode to reduce spattering. Generally, good weld beads can be produced at high MIG arc currents, as the situations of laser welding and hybrid welding of an aluminum alloy are schematically shown in Fig. 6.21 [20, 21]. Porosity is formed in laser welding and hybrid welding at low MIG arc currents, but such porosity can be reduced or prevented at high arc currents. The prevention of porosity is attributed to no formation of bubbles under the defocused condition of a laser beam due to the concave molten pool surface and the disappearance of bubbles from the concave molten pool made by high arc currents when the keyhole is not so deep (if bubbles are formed).

Subsequently, weld penetrations, surface appearances, and porosity formation tendency in YAG-MIG and MIG-YAG hybrid welding of A5052 aluminum alloy at the arc current of 240 A are compared at various target distances of an arc and a laser beam in Fig. 6.22 [20, 21]. In YAG-MIG hybrid welding, sound weld beads of good surface appearances can be produced by cleaning action due to MIG arc. The deep weld bead can be formed at the proper distance between a laser spot and arc target location. In MIG-YAG hybrid welding, deeper weld beads can be produced

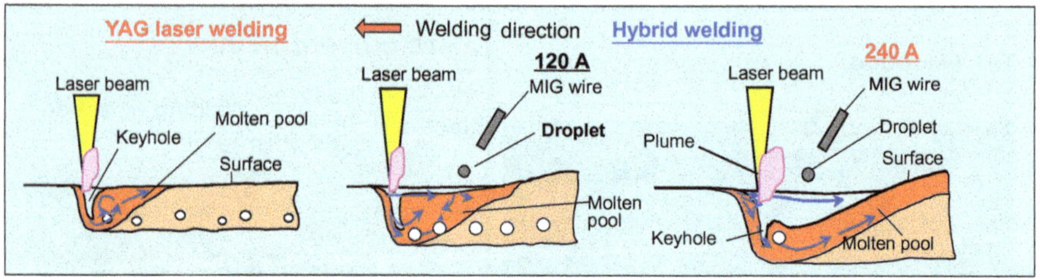

Fig. 6.21 Schematic situations of laser welding and laser–arc hybrid welding of aluminum alloy at arc current of 120 A and 240 A

Distance, *d* [mm]	YAG-MIG			MIG-YAG		
	Surface	Cross section	X-ray inspection	Surface	Cross section	X-ray inspection
0						
2						
4						Porosity
6						
8						
10						

Fig. 6.22 Surface appearances, cross sections, and X-ray inspection results of YAG-MIG and MIG - YAG. hybrid welds in A 5052 aluminum alloy at various laser–arc target distances

at the proper laser–arc distances, but the weld bead surfaces are dirty because they are covered with evaporated and oxidized particles and porosity is present in the deep weld beads. The reason for deeper weld penetration is attributed to the laser irradiation on the molten pool made by MIG arc. On the other hand, bubbles are easily formed with the following laser beam, resulting in the porosity formation. The porosity in MIG-YAG hybrid welding can be prevented in deep weld beads by using the forward-inclined laser beam. Hybrid welding with YAG, disk, and fiber laser and MIG arc in this arrangement is applied to weld aluminum alloy sheets in the automotive industry.

In the hybrid welding of steels with YAG laser and MAG arc, pulsed MAG arc was recommended to suppress spattering from short circuits of the arc. Moreover, cold metal transfer (CMT) welding is recommended for the reduction in spattering. Deeper weld penetration can be stably obtained in MAG-(YAG, disk or fiber) laser hybrid welding (hybrid welding of the leading MAG arc and the following laser) than laser-MAG hybrid welding. In hybrid welding of thick steel plates, MAG arc and disk or fiber laser hybrid welding are generally performed. Nevertheless, it is important to select the proper welding conditions, especially proper arc currents, as shown in Fig. 6.23 [22]. When the arc current is slightly low, weld beads with humps on the bottom surface are easily formed. When the arc current is too high, underfilled beads are formed. The proper arc current should be found out.

In welding of steels, CO_2 gas arc and disk or fiber laser hybrid welding was performed. As the hybrid welding results are shown in Fig. 6.24 [22], sound deep welds without porosity can be satisfactorily produced under the wider welding conditions than MAG-laser hybrid welding. Spattering is slightly violent when CO_2 gas arc welding is used, but this spattering can be greatly reduced by the selection of buried arc conditions and the use of pulsed CO_2 gas arc machine developed recently if possible. Hybrid welding with MAG or CO_2 gas arc and disk or fiber laser in this arrangement has been applied to weld thick steel plates in the shipbuilding industry.

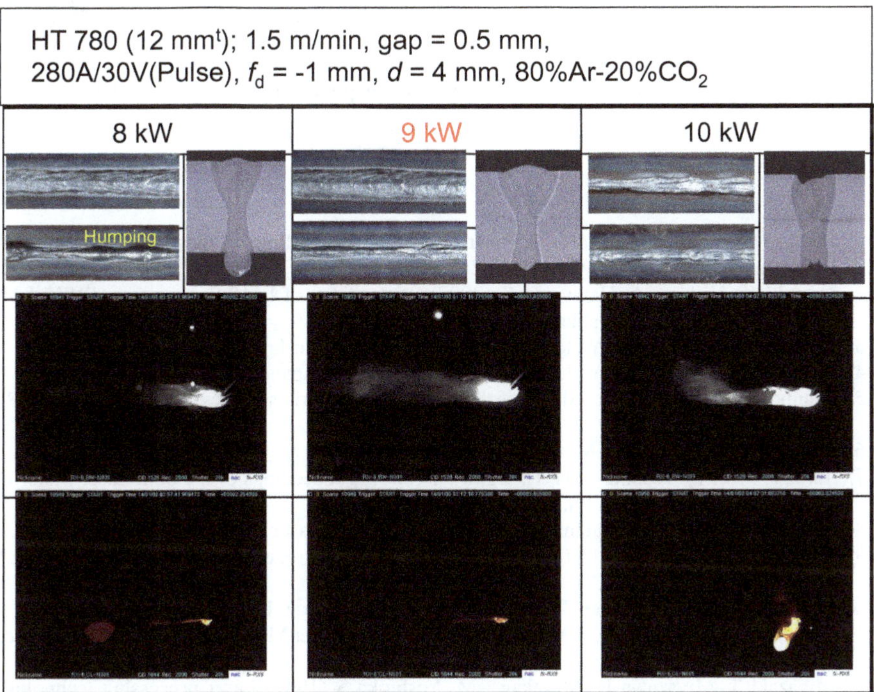

Fig. 6.23 Top and bottom surface appearances and cross sections of hybrid welds, and high-speed video observation results in two directions during MAG arc–disk laser hybrid welding

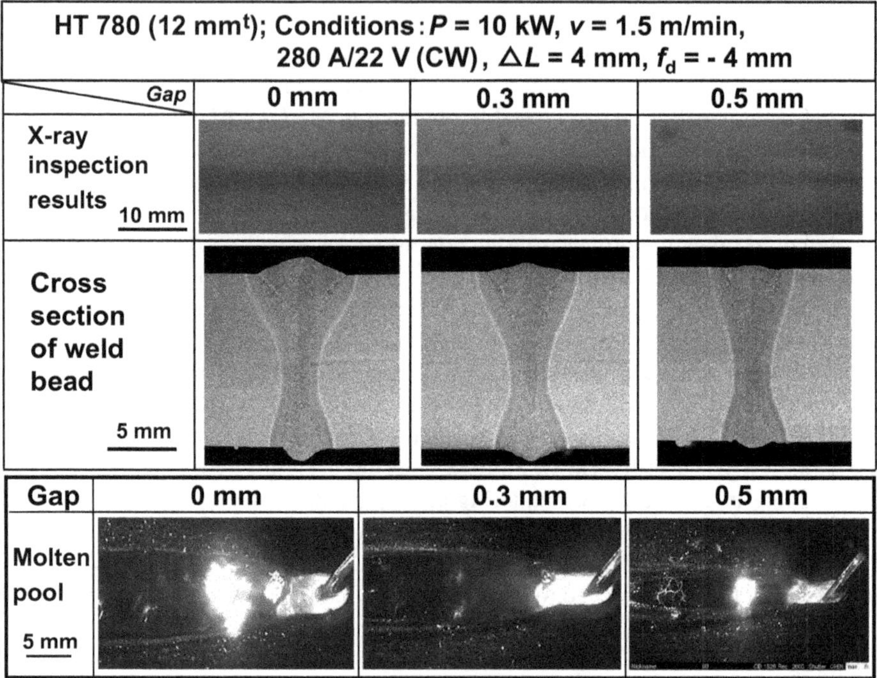

Fig. 6.24 X-ray inspection results and cross sections of hybrid welds with CO_2 gas arc and disk laser, and high-speed video photographs showing molten pool, buried wire, and keyhole

References

1. Pic A, Nunera DD, Vierstraete R, Pinard F (2006) Proceedings of the 67th Laser Materials Processing Conference, JLPS, pp 49–62
2. Oowaki K, Nishimi A, Hatayama K (2007) Proceedings of LMP Symposium, JWES LMP committee, pp 13–18 (in Japanese)
3. Katayama S (2015) J Jpn Weld Soc JWS 84(8):582–590 (in Japanese)
4. Katayama S (2020) J Jpn Weld Soc JWS 89(1):5–15 (in Japanese)
5. Naumovski V (2007) Proceedings of the 68th Laser Materials Processing Conference, JLPS, pp 151–155
6. Mikame K (2009) J Laser Mater Process JLPS 16(1):8–15 (in Japanese)
7. Tarui T, Mori K, Yoshikawa H, Hasegawa T (2007) Proceedings of the 68th Laser Materials Processing Conference, JLPS, pp 157–163 (in Japanese)
8. Oiwa S, Kawahito Y, Mizutani M, Katayama S (2011) J Laser Appl 23(2):022007-1-7
9. Kimura S, Takemura S, Mizutani M, Katayama S (2006) Proceedings of ICALEO 2006, LIA, Paper#528, pp 346–354 (CD)
10. Beyer E (2018) (Fraunhofer IWS), Personal Communication, Laser Market Place 2003
11. Yamamoto S, Kishino S (2010) Proceedings of the 73rd Laser Materials Processing Conference, JLPS, pp 57–61 (in Japanese)
12. Miyazaki T, Miyazawa H, Murakawa M, Yoshioka S (1991) Laser processing technology. Sangyo-Tosho (Industry-Library), Ltd., p 18 (in Japanese)
13. Nippon Avionics Co., Ltd. (2018) Web Homepage. http://www.avio.co.jp/products/assem/application/electronic-parts/soldering

14. Katayama S, Hirayama M, Mizutani M, Kawahito Y (2012) 65th Annual Assembly of Int. Inst. Welding (IIW), IIW Doc. IV-1097-12/XII-2096-12/212-1256-12
15. Moon J-H, Mizutani M, Katayama S, Matsunawa A (2003) J Laser Appl, LIA 15(1):37–42
16. Takeda S (2019) Personal communication. Laserline 14. Web https://www.laserline.com/
17. Data presented from Mitsui Bussan Electronics Ltd. and NADEX Co., Ltd.; Data of Corelase (Coherent; Rofin-Sinar) Laser
18. Naito Y, Mizutani M, Katayama S (2006) J Jpn Weld Soc JWS 24(2):149–161 (in Japanese)
19. Naito Y, Mizutani M, Katayama S (2006) J Laser Appl LIA 18(1):21–27
20. Uchiumi S, Wang JB, Katayama S, Mizutani M, Hongu T, Fujii K (2004) Proceedings of ICALEO 2001, LIA, Florida, USA, P 530 (CD)
21. Katayama S, Naito Y, Uchiumi S, Mizutani M (2005) Lasers in Manufacturing 2005, WLT, pp 193–198
22. Pan QL, Mizutani M, Kawahito Y, Katayama S (2016) J Laser Appl LIA 28(No. 1):012004-1-9

Chapter 7
Process Monitoring, Sensing, and/or Adaptive Control during Laser Welding

7.1 Process Monitoring Technology

To produce high-quality welds, sensing, monitoring, and feedback or adaptive control are necessary and have been intensively investigated. Monitoring during welding should be satisfactorily understood by dividing three processes: pre-process monitoring or sensing, in-process or online monitoring, and post-process monitoring, as shown in Fig. 7.1 [1]. The location to be welded can be detected by the sensing technique beforehand as pre-process monitoring, and weld-line tracking technology in seam-welding according to the detected location has been developed. Situations during welding are detected and the presence or absence of welding defects and the good or bad quality of a weld are evaluated as in-process monitoring. It is ideal to always produce sounds welds by preventing welding defects under the adaptive control based upon the in-process monitoring during welding. In the post-process, the surface contours of weld beads are measured to detect the presence or absence and the degree of surface welding defects such as underfills, undercuts, and pits.

7.2 Sensing or Seam Tracking during Laser Welding

At present, the apparatuses capable of sensing, tracking, and post-process monitoring have been developed. An example of the apparatus and measurement results are shown in Fig. 7.2 [2]. And the apparatuses and systems capable of sensing, tracking, and/or keyhole depth measurement as in-process monitoring and post-process monitoring of weld bead surface geometry have been developed. The welding situations and the monitoring results are shown in Fig. 7.3 [3]. Sound welds can be produced by this system of weld-line tracking.

A large number of studies have been actively devoted to developing the system of in-process monitoring capable of detecting the occurrence of welding defects.

S. Katayama, *Fundamentals and Details of Laser Welding*,
Topics in Mining, Metallurgy and Materials Engineering,
https://doi.org/10.1007/978-981-15-7933-2_7

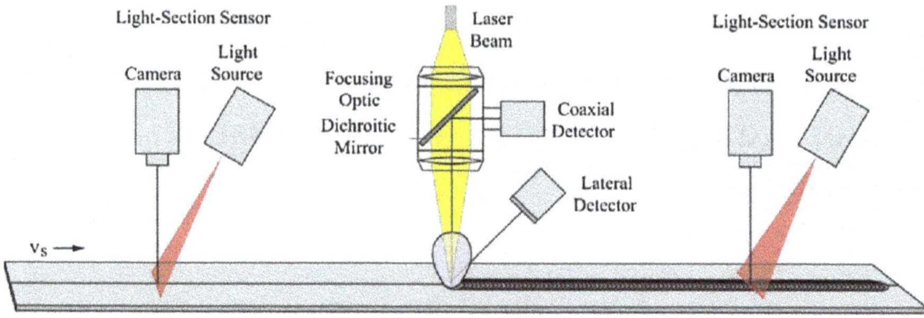

Fig. 7.1 Schematic illustration of pre-process monitoring or sensing, in-process or online monitoring and post-process monitoring for laser welding

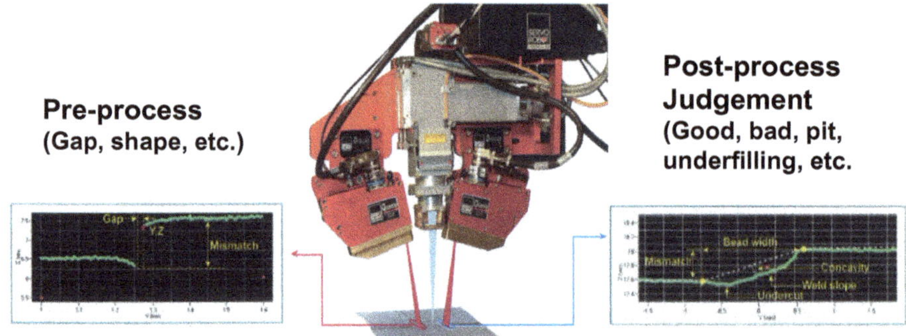

Fig. 7.2 Example of apparatus capable of seam tracking and measurement results of pre-process and post-process

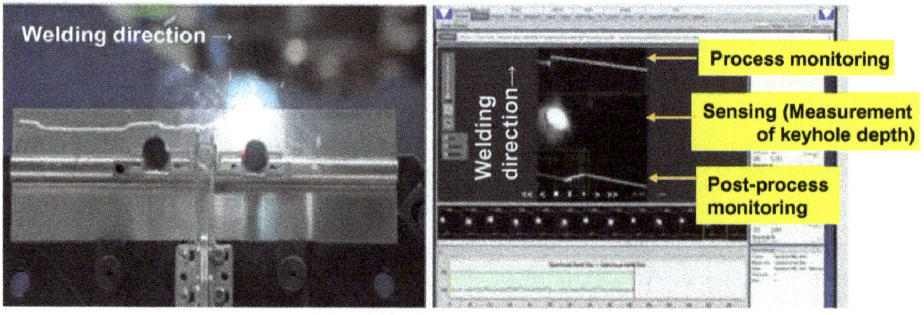

Fig. 7.3 Laser welding situation during monitoring and display of monitoring results of pre-, in- and post-process during welding

As in-process monitoring signals during laser welding, a reflected laser beam, light emission, and spectroscopic lines from a plasma or a plume, thermal radiation and temperature profiles from a molten pool, a sound from the plasma formation or keyhole inlet, ultrasonic or acoustic sound from the metal inside, plasma potential between the plate and the nozzle (in the case of CO_2 laser welding) or laser-induced plume, and so on are investigated with conjunction with weld penetration or welding defects. The reflected laser beams and thermal radiation signals are used in the actual production lines. An example of online technique used in the actual production line of aircraft panels of A6XXX alloy with a filler wire including about 12%Si is schematically illustrated in Fig. 7.4 [4]. The emission lines from Si atoms, meaning the melting of a filler wire, are detected to judge sound welds without solidification cracks or poor welds with solidification cracks, because the weld fusion zone including high amount of Si is resistant to solidification cracking in aluminum alloys. Coaxial imaging observation can judge a gap and a difference between full and partial weld penetration in sheets or plates. The other laser beams are also used as a source to measure the keyhole depth and to detect the presence of welding defects.

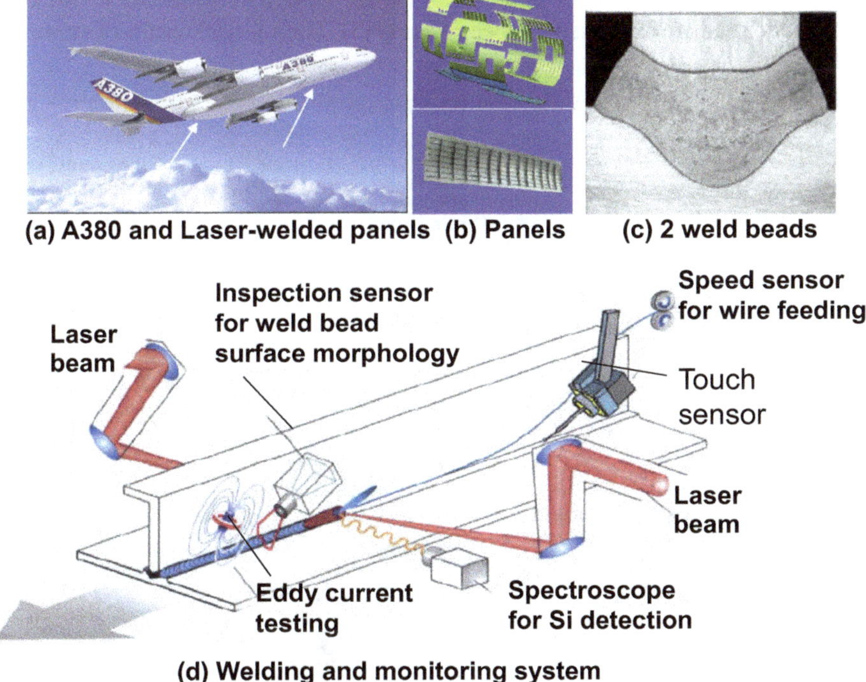

Fig. 7.4 Example of airplane adopting panels made by laser welding, cross section of laser-welded joint, panel with stringers, and schematic illustration of online technique used in actual production line of aircraft panels of A6XXX alloy with filler wire

7.3 OCT and Its Application to Keyhole Depth Measurement

Presently, optical coherence tomography (OCT) is noted as imaging technology making a good use of coherent light. The OCT technology can capture two- and three-dimensional images at the resolution of 1 μm from the object capable of optical reflection and scattering. The OCT devices which can measure a keyhole depth during laser welding have been developed in Germany and Canada in around 2010 and are now commercially available. A special laser beam of about 0.85 μm or 1.5 μm in wavelength, which is different from the wavelength of a laser beam for welding, is utilized for OCT system to detect the reflection lights from the plate surface and keyhole tip leading to the keyhole depth. The measurement procedure of the keyhole depth is schematically illustrated in Fig. 7.5 [3, 5, 6]. A low coherent light beam is emitted from a broadband light source (x). It is divided into two light beams of a reference light beam (1) and a measuring light beam (2) by the beam splitter. Reflected light beams (3) from the reference plane (1) and the sample surface (4) or the keyhole tip are, respectively, synthesized, and consequently, their interference occurs. The interference light beams (3) are guided to the sensor by the beam splitter and then are dispersed by spectroscopy (5). The spectral results of strong and

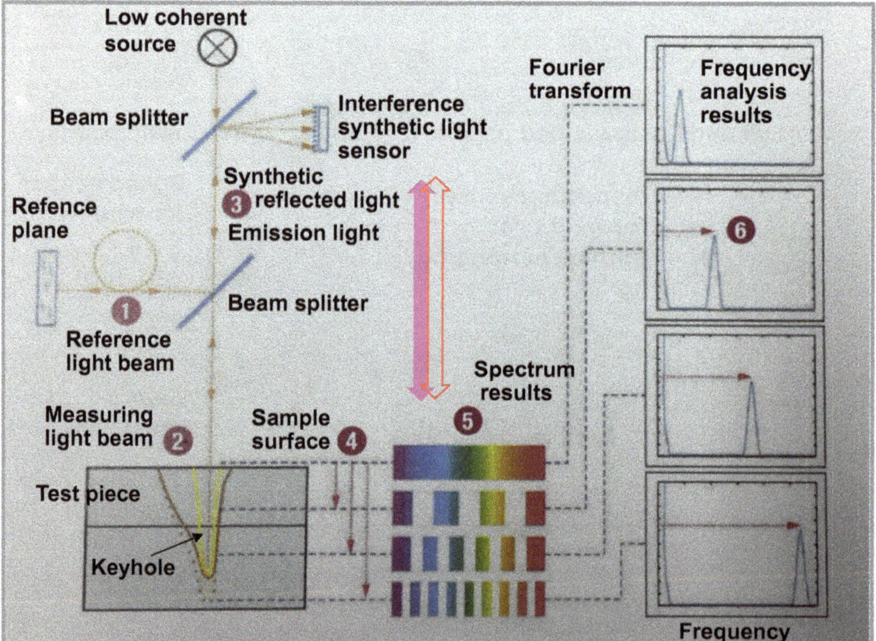

Fig. 7.5 Optical coherence tomography (OCT) measurement procedures of keyhole depth during laser welding

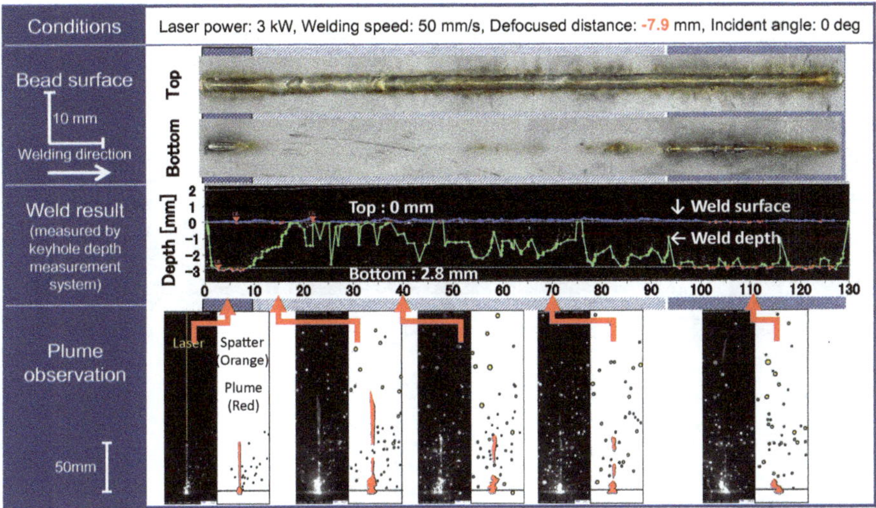

Fig. 7.6 Top and bottom surfaces of laser weld beads in Type 304 steel sheet of 3 mm thickness, plume and spattering situations during welding and OCT measurement results in welding in air without shielding gas, showing unstable penetration depths of melt-run weld

weak electromagnetic waves are subjected to Fourier transform (6). The frequency becomes high with an increase in the keyhole depth (4). As a result, the depths of a sample surface and a keyhole tip can be measured from the respective frequencies. Finally, actual keyhole depths and their variation can be displayed as the difference of respective depths.

In laser welding, OCT apparatuses are used. The welding results of the top and bottom surfaces of laser weld beads in Type 304 steel sheet of 3 mm thickness, plume situations during welding and OCT measurement results without and with a shielding gas are shown in Figs. 7.6 and 7.7 [7]. Deep laser welds are stably produced in the shielding gas. It is therefore judged that OCT results display the weld penetration results. Such results are confirmed by comparison of the penetration depths in the longitudinal cross sections with the OCT results in many researches. It is also confirmed that OCT measurement results of keyhole depths are in good agreement with the measurement results of keyholes observed by X-ray transmission imaging system [7]. Moreover, the feedback mechanism of laser powers based on the OCT results is developed to keep the constant keyhole depth and to produce a weld of desired penetration depth. This result can be confirmed in Fig. 7.8 [6]. A difference in the depth between laser weld beads made without and with OCT and control system is noticeable.

The OCT system capable of measuring and sometimes contoling keyhole depth during laser welding is extremely beneficial and important to know the penetration depth and to secure welding depths.

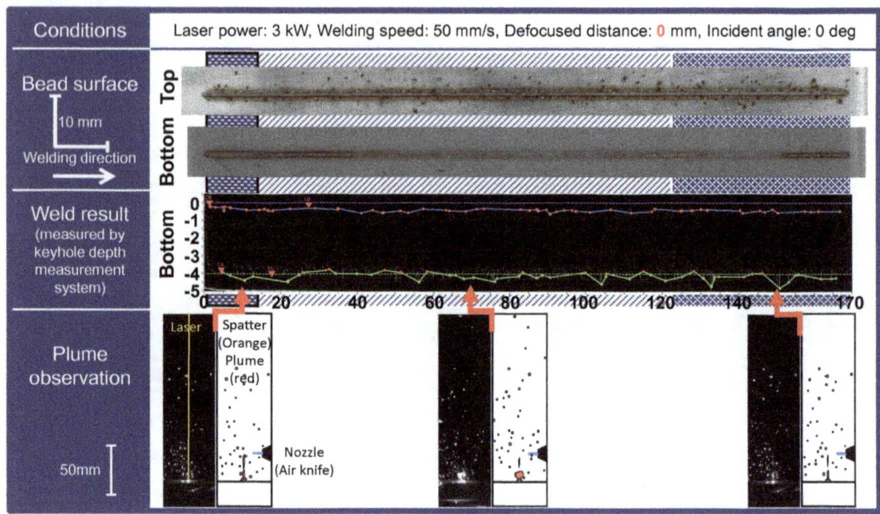

Fig. 7.7 Top and bottom surfaces of laser weld beads in Type 304 steel sheet of 3 mm thickness, plume and spattering situations during welding and OCT measurement results in welding with shielding gas, showing decreased plume height and stable penetration depths of melt-run weld

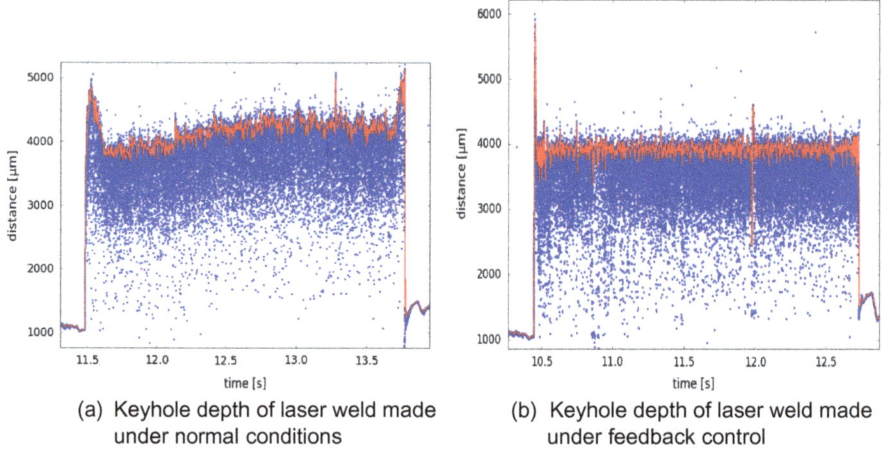

Fig. 7.8 OCT results of keyhole depths during laser welding under normal conditions and controlled conditions to produce desired penetration depth

7.4 In-Process Monitoring and Adaptive Control during Pulsed or Continuous Wave Laser Welding

Pulsed YAG lasers of low heat inputs and disk and fiber lasers of high beam-focusing capability are used for welding of fine precision parts. These welding results are affected by slight differences of materials surface conditions, jigs and the pressing degree, the gap between sheets, and so on, sometimes resulting in the occurrence of defective products or inferior goods. Thus, the adaptive controllable YAG laser apparatuses are developed as devices of the highest intelligence. In particular, the in-process monitoring system capable of evaluating good or poor products based upon the laser power, and reflected light and thermal radiation have been developed, as the system is schematically presented in Fig. 7.9 [8, 9]. The adaptive control system capable of in-process repairing, which can diagnose the welding processes in microsecond (ms) by using thermal radiation values during welding with a pulsed laser of several millisecond (ms) width and is developed to always produce sound welds.

In fact, in the case of a thin upper sheet and welding-difficult materials such as aluminum (Al) alloys and copper (Cu), hole defects are easily formed or non-lap-welding occurs because the upper sheet is readily deformed. Laser welding phenomenon is detected by using the laser apparatus, in-process monitoring, and adaptive (feedback) control system, and consequently, spattering can be suppressed by decreasing laser power when the spattering is about to happen soon. Moreover, when no lap joining of the upper and lower sheets due to the deformed upper sheet is judged by the adaptive control system, the laser power is increased to lap-weld two sheets by melting the lower sheet, as the monitoring results of laser power, reflected laser and thermal radiation and the adaptive repairing results during pulsed YAG laser spot welding are shown in Fig. 7.10 and schematic illustration of in-process repairing mechanism is indicated in Fig. 7.11 [8, 9]. Hole defects can be suppressed or prevented stably producing an enough large molten pool in spot welding with a pulsed YAG laser, resulting in the formation of sound welds at all times. Such welding with adaptive control system is concluded to be an ideal welding process owing to joining without failure.

In CW laser welding, it is important and required to produce a weld bead with the same bead width or the same penetration depth. The weld bead width may be affected because of the reduction in thermal conductivity due to the presence of a

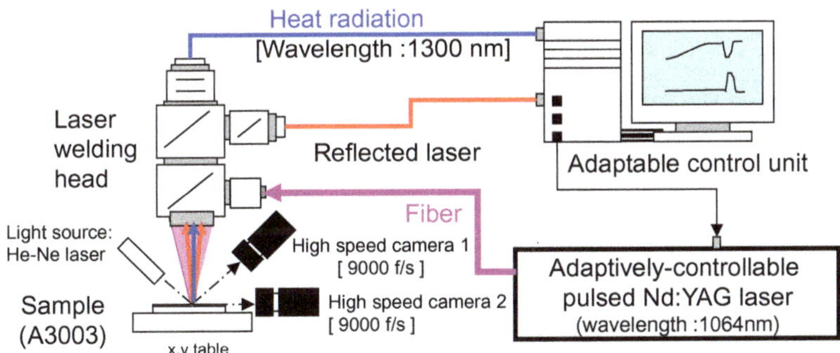

Fig. 7.9 Schematic arrangement of in-process monitoring and adaptive controllable apparatus capable of measuring laser power, reflected light, and thermal radiation

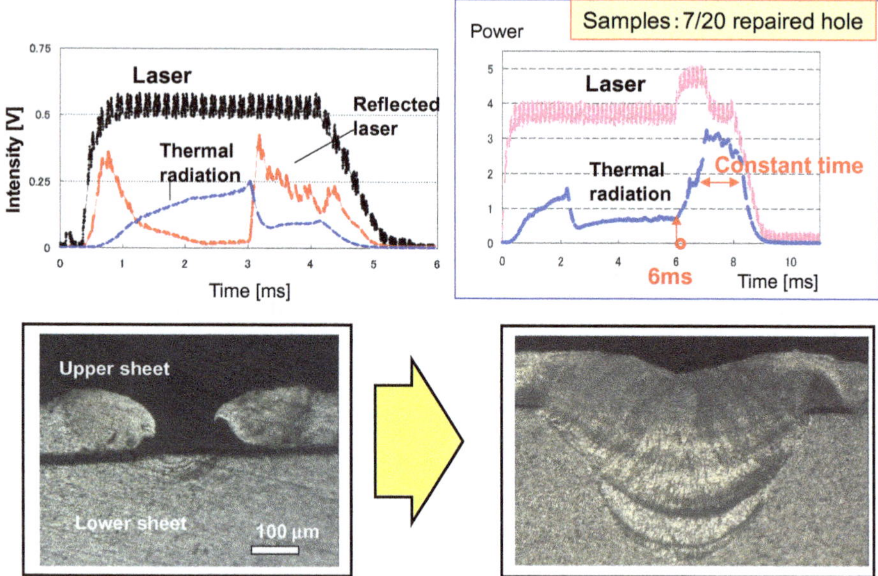

Fig. 7.10 Comparison of pulse shapes, thermal radiation signals, and cross-sectional welds produced by normal welding and by in-process repairing of lap joints in pulsed YAG laser spot welding of A3003 aluminum alloy sheets

hole in the lower jig plate in bead welding with a CW fiber laser of a thin sheet, as shown in Fig. 7.12 [10]. The width variation can be monitored by the signal values of thermal radiation during welding. It is consequently confirmed that the control of the laser power on the basis of the thermal radiation values can minimize the variation in the bead width. It is moreover demonstrated that, in deep penetration welding, the weld penetration can be controlled by using OCT system capable of monitoring the keyhole depth and adjusting the laser power, as the results were shown in Fig. 7.8 and described in Sect. 7.3 [6].

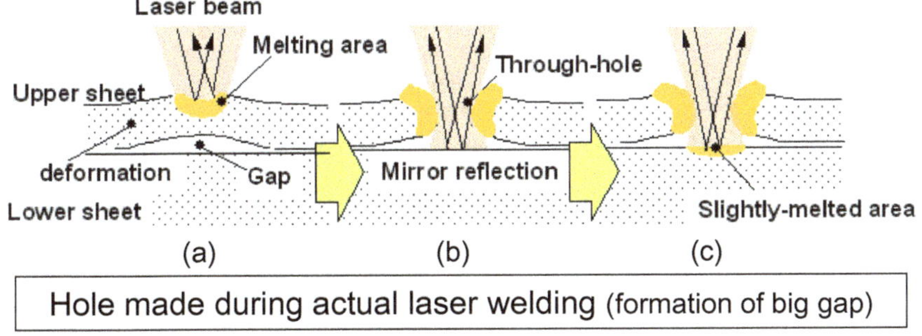

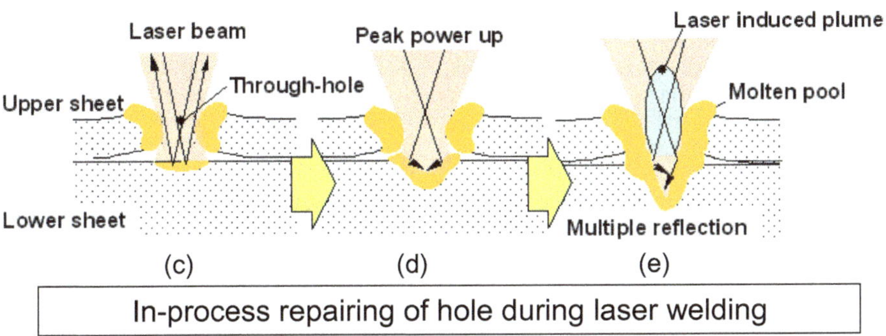

Fig. 7.11 Schematic illustration of hole formation in normal welding and in-process repairing mechanism of pulsed YAG laser spot welding of lap joint in A3003 aluminum alloy sheets

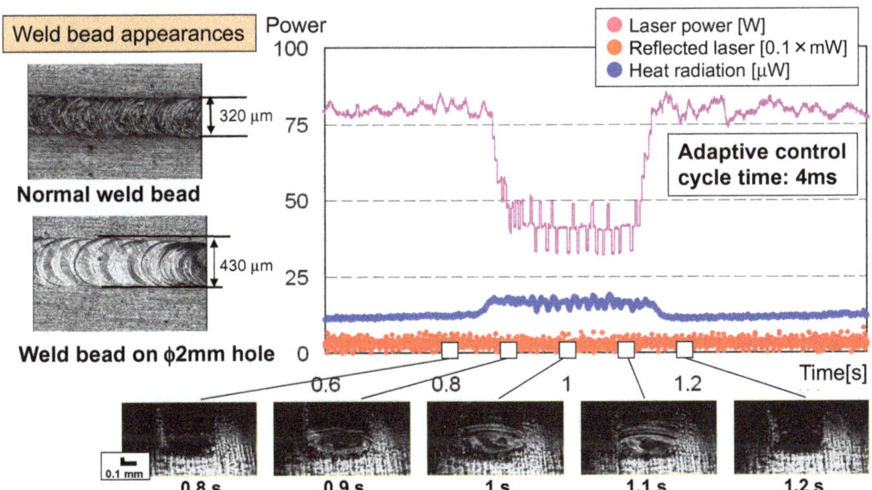

Fig. 7.12 Feedback-controlled laser power, reflected laser power, and heat radiation for keeping constant bead width of laser weld in sheets, observation results of molten pool during welding, and weld beads made without and under control

References

1. Petereit J, Abels P, Kaierle S, Kratzsch C, Kreutz EW (2002) Proceedings of ICALEO 2002 (Laser Materials Processing Conference), LIA, Session A-Welding (CD)
2. SERVO-ROBOT (3D Robot Vision Systems) Pamphlet, (Web Homepage; https://servo-robot.com)
3. Precitec Japan Pamphlet, (Web Homepage; https://www.precitec.de/jp/Precitec-group-start-page/) (Laser Photonics Exhibition 2017)
4. Schumacher J, Zerner I, Neye G, Thormann K (2002) Proceedings of ICALEO 2002 (Laser Materials Processing Conference), LIA, 94 Session A-Welding (CD)
5. Bautze T, Moser R, Strebel M, Kogel-Hollacher M (2015) Proceedings of LiM 2015, WLT, Macro Processing: Process Monitoring and Control Session, pp. 1–9 (Memory Stick)
6. Kogel-Hollacher M (2015) Proc. of the 83rd Laser Materials Processing Conference, JLPS, pp 83–87 (in Japanese)
7. Sumimori D, Deguchi T, Nomura R, Kamei N, Katayama S, Mizutani M, Kawahito Y (2016) Reprint of the Japan Welding Society Annual Meeting, JWS (In Japanese)
8. Kawahito Y, Katayama S (2004) The review of laser engineering. Laser Soc Japan 32(5):357–363 (in Japanese)
9. Kawahito Y, Katayama S (2005) J Laser Appl LIA 17(1):30–37
10. Kawahito Y, Kawasaki M, Katayama S (2008) JLMN J Laser Micro/Nanoeng 3(1):46–51

Chapter 8
Features of Laser Welding or Joining of Various Materials

8.1 Laser Welding of Steels or Stainless Steels

Steels are very well-known metals used mostly as our surroundings such as cars, trains, ships, bridges, architecture, machinery, and so on. They are variously classified and called: mild steel, dead soft steel (from hardness), low carbon steel, high carbon steel, low alloy steel, Ni steel, Cr–Mo steel (from component), high tensile strength (HT) steel, ultrahigh strength steel (from strength), heat-resistant steel, low temperature steel, stainless steel (from quality or properties), cold rolled steel, hot rolled steel, cast steel (from manufacturing process), quenched and tempered steel, maraging steel (from heat treatment), surface treatment steel, Zn-coated steel (from post-treatment), converter steel, electric furnace steel (from steelmaking), rimmed steel, killed steel (from deoxidation method), ferrite-perlite steel, bainitic steel, martensitic steel, dual phase steel (from microstructure), and automotive steel, structural steel, pressure vessel steel, tool steel (from use). They are also named general structural rolled steel (SS; SS400), welding structural rolled steel (rolled steel for welded structure) (SM; SM490A), carbon steel for machine structural use (SC; S45C), low alloy steel for machine structural use, and high strength steel (590, 780 MPa, etc.) in Japanese Industrial Standards.

Steel sheets and plates have been subjected to laser welding in Ar, He, or N_2 shielding gas. Steels are generally regarded as easily laser-weldable materials except for high carbon (C) steel. Nevertheless, porosity and cracking may take place. Porosity should be formed from the bubbles generated from a keyhole tip during a keyhole type of laser partial penetration welding at the low welding speeds of less than 60 mm/s (3 m/min) in Ar or He shielding gas. Such porosity can be reduced in full-penetration welding at any speeds or in partial penetration welding at higher welding speeds. Besides, porosity can be drastically reduced by using 100% CO_2 shielding gas for steels in welding with YAG, disk, or fiber laser, although porosity may be formed in CO_2 laser welding at high powers in 100% CO_2 shielding gas. Porosity can be also reduced in welding in vacuum. It is also noted that hot cracking

S. Katayama, *Fundamentals and Details of Laser Welding*,
Topics in Mining, Metallurgy and Materials Engineering,
https://doi.org/10.1007/978-981-15-7933-2_8

or solidification cracking may occur in partial penetration welds of thick steel plates, as already shown in Figs. 3.20 and 5.13. This is attributed to delayed solidification of a wide retained liquid area in the lower part of a weld bead, as already described in Sect. 5.3. In the case of high carbon steels, solidification cracking is likely to occur in grain boundaries due to the primary solidification of austenite because of a higher content of C (Carbon) and microsegregation of impurities such as S (sulfur) and P (phosphorus), and cold cracking may occur due to the full formation of hard martensite and cementite phases. Such cracking may be present in laser welds of high strength steels. In laser welding, such cold cracking may propagate in the laser welds from the origin of solidification cracks in the crater of the final solidification location.

In almost all steels except for HT steels of 780 MPa or higher, the hardness of laser welds is higher because of a higher amount of bainite and martensite phase due to rapid cooling than that of the base metal. An example of hardness profile across a weld bead, the macro-cross section of laser weld and the microstructures of laser weld fusion zone and near the fusion boundary of HT590 steel are shown in Fig. 8.1 [1]. A harder laser weld is confirmed in HT590. On the other hand, 1500 MPa HT

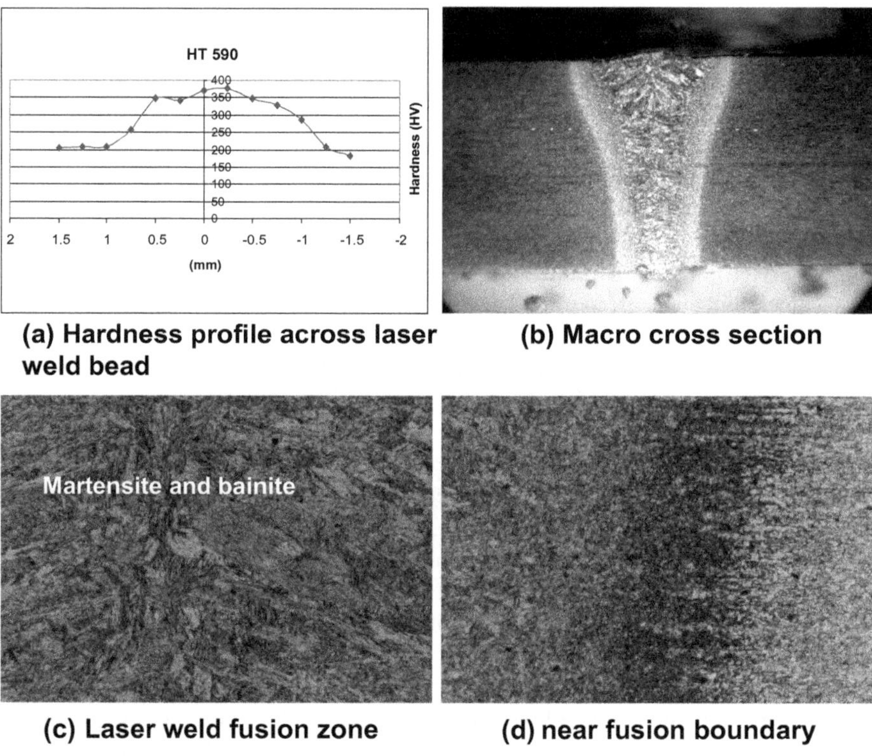

Fig. 8.1 Example of hardness profile across laser weld, macro-cross section of laser weld bead, and microstructures of laser weld fusion zone and near fusion boundary of HT590 steel

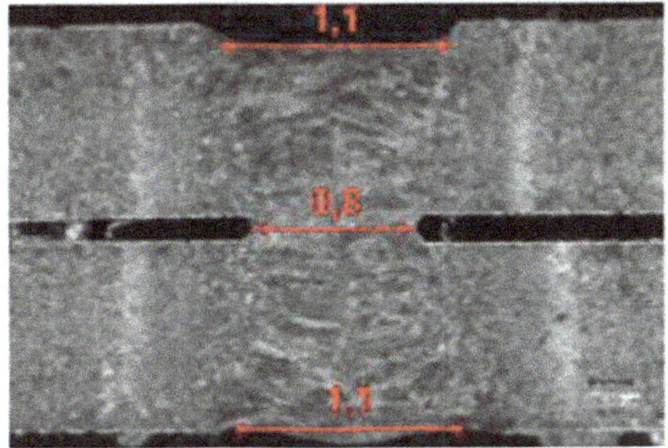

(a) Cross section of HT1500 steel

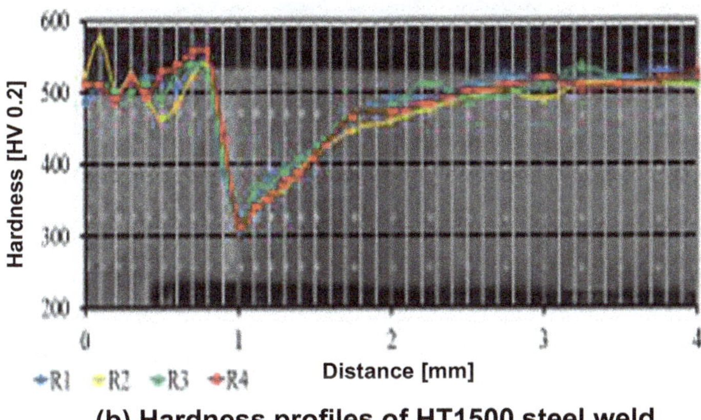

(b) Hardness profiles of HT1500 steel weld

Fig. 8.2 Cross section of 1500 MPa HT steel weld and hardness profiles, showing HAZ softening

steel becomes soft in the HAZ near the base metal, as the weld and hardness profiles are shown in Fig. 8.2 [2]. This HAZ softening is attributed to the formation of soft ferrite phase during heating and cooling history. (In actually, perlite phases of a large amount of soft ferrite and a small amount of hard cementite are formed, and soft HAZ is produced by the formation of a large amount of soft ferrite phase.) These points should be noted for the actual welded products and in laser tailored blank welding of such ultrahigh strength steels.

In lap welding of Zn-coated steel sheets with a laser beam, a gap of about 0.1–0.2 mm between sheets should be set to prevent spattering and porosity, as already described in Sect. 5.2. Similarly, ZAM-coated steel lap sheets can be welded with fiber or disk laser. Laser brazing is also commercially used to joint such steel sheets in the automobile industry.

In the case of cast iron, quenching cracking may take place in cementite phase due to rapid cooling. To prevent such cracking, a wire with a high content of nickel

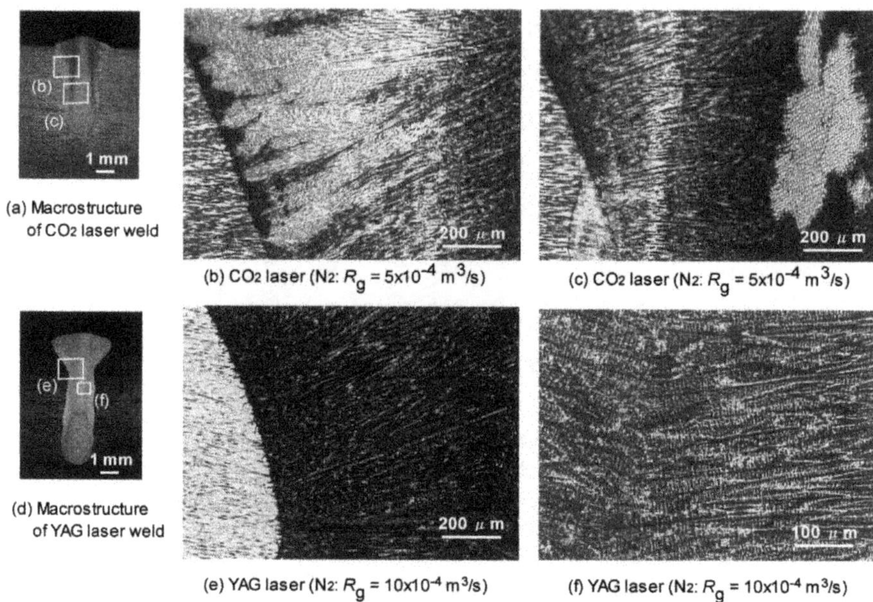

Fig. 8.3 Comparison of room temperature microstructures of weld metals in stainless steels made with CW CO_2 and YAG lasers in nitrogen shielding gas

(Ni) may be used to facilitate the formation of softer ductile austenite phase. It is also important and necessary to select proper welding conditions.

In laser welding of austenitic stainless steels, N_2 shielding gas is preferably recommended for YAG, disk, fiber, and diode laser welding except for CO_2 laser welding at low speeds. The X-ray transmission imaging photos during YAG laser welding in Ar, He, and N_2 shielding gas are already compared in Fig. 5.7. Bubbles and porosity can be reduced in N_2 shielding gas. The weld microstructures of YAG and CO_2 laser welds are compared in Fig. 8.3 [3]. In YAG laser welds, the primary delta (δ)-ferrite phase solidification takes place and the ferrite content is reduced during cooling. In CO_2 laser welding, the absorption of N takes place through a keyhole during welding, and the N-enriched area is solidified as primary austenite phase and solidification cracking sometimes occurs along austenite grain boundaries enriched with microsegregation of P and S impurities. The formation of N plasma should enhance N absorption in the weld fusion zone made with CO_2 laser.

The microstructure of CW laser weld metals in stainless steels at room temperature except for high welding speeds of more than 10 m/min is observed to be similar to that of arc welds. It should be noted that the room temperature microstructure of stainless steels is different from the high temperature one during solidification, as the microstructures during solidification and the subsequent high-temperature transformation process are schematically illustrated in Fig. 8.4 [4]. Various austenitic and ferritic microstructures are observed in weld fusion zones depending upon the chemical compositions of stainless steels. The content of δ-ferrite should be largely

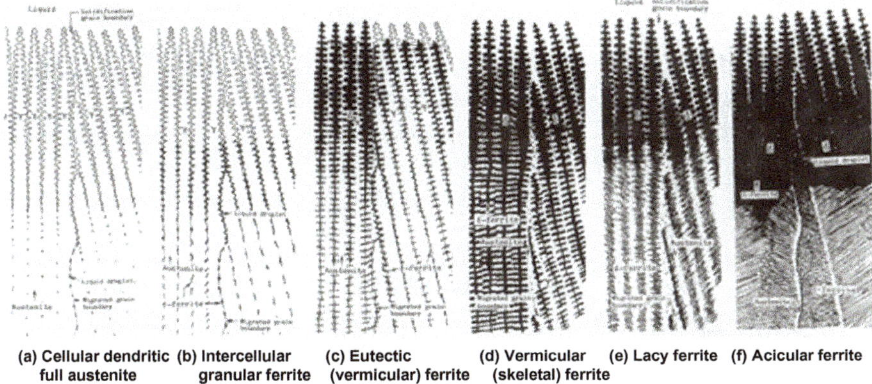

Fig. 8.4 Schematic illustration of typical ferrite and austenite microstructures of weld metals during solidification of different stainless steels

reduced at room temperature by transformation. Nevertheless, the ferrite contents of CW laser weld fusion zones except for those made at high welding speeds may be predicted from the Schaeffler diagram, Delong diagram, etc.

The microstructures and ferrite contents of pulsed laser weld metals at room temperature are greatly different from those of arc weld fusion zones or high-power CW laser weld metals, as shown in Fig. 8.5 [5]. The residual δ-ferrite contents are indicated on the Schaeffler diagram, but the full austenite and full ferrite regions are expanded, resulting in narrow duplex austenite and ferrite region in pulsed spot weld metals subjected to rapid solidification and rapid cooling. For example, Type 304 weld is fully austenitic in pulsed laser spot weld metal, but solidification cracking hardly occurs, as the microstructures of fully austenitic Type 310S steel and Type

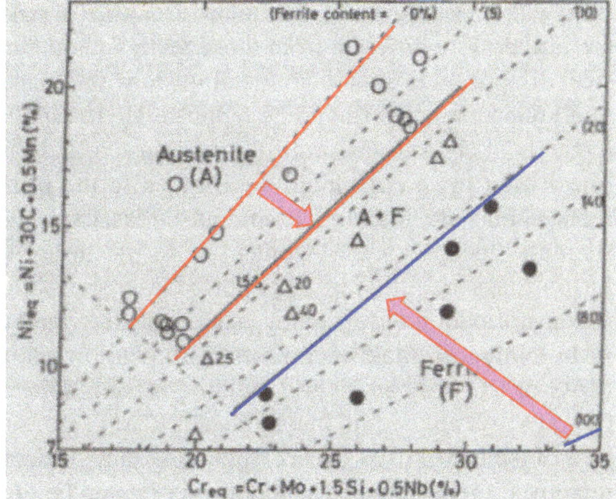

Fig. 8.5 Ferrite contents of pulsed YAG laser spot weld metals at room temperature projected on Schaeffler diagram, showing narrowness of ferrite-austenite dual phase region due to rapid solidification and rapid cooling

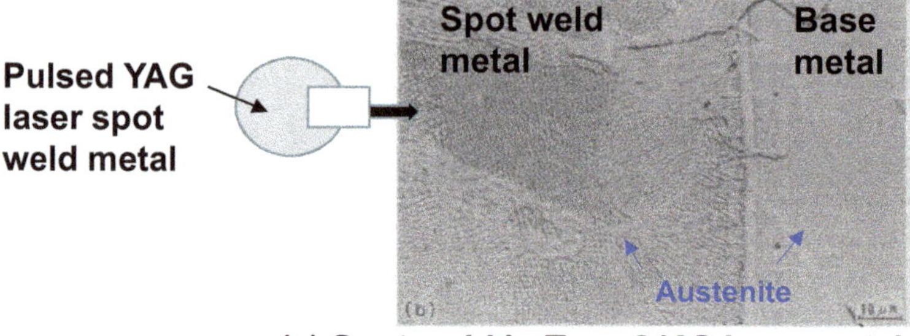

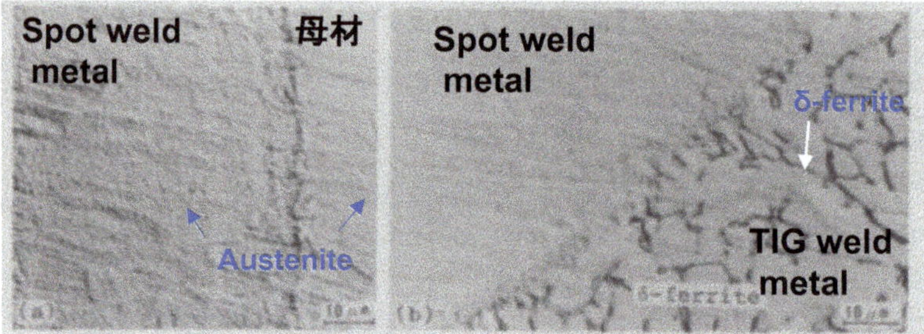

Fig. 8.6 Microstructures of pulsed YAG laser spot weld fusion zones of Type 310S (**a**) and Type 304 steels in comparison with those of base metal (**b**) and TIG weld fusion zone (**c**), showing reduced δ-ferrite content in spot weld metal due to rapid solidification and rapid cooling

304 are shown in Fig. 8.6 [5]. Extremely fine cells are observed in Type 310S wed fusion zone, meaning rapid solidification. In the case of Type 304 steel, spot welds are produced in as-received base metal with a negligible content of ferrite and TIG arc weld with about 5% ferrite. The microstructures of pulsed laser spot weld fusion zones are furthermore different from those at high temperatures, and δ-ferrite in Type 304 steel disappears in spot weld metal made with pulsed YAG laser. Microsegregation of alloying elements should be reduced due to rapid solidification, and the primary ferrite is likely to transform fully into austenite during cooling after solidification. The reason for no cracking is attributed to the primary ferrite solidification with reduced microsegregation during pulsed laser spot welding. On the other hand, the stainless steels with 20% content of ferrite in arc or high-power CW laser weld fusion zones are fully ferritic in pulsed laser spot weld metals due to the suppression of transformation of ferrite to austenite during rapid quenching [6].

In joining of thick steel plates in shipbuilding, currently, disk or fiber laser and MAG or CO_2 gas arc hybrid welding is used. Generally, in the case of the use of pulsed MAG arc or cold metal transfer (CMT) arc, spattering can be reduced. In the case of the use of CO_2 gas arc, spattering may be decreased by using buried arc, and moreover, the weld metal without porosity can be produced.

8.2 Laser Welding of Aluminum Alloys

Aluminum (Al) and its alloys have excellent properties of the specific gravity of 2.7 meaning approximately one-third times as light as steel, superior plastic workability, good corrosion resistance in air and water due to surface oxide films, and so on. Thus, they are used for cars, fast and light vehicle, ship, aircraft, aerospace equipment, civil engineering, architecture, battery cases, and so on. Al alloys are divided broadly into rolled material (processed alloy) and cast material (casting alloy). Moreover, they are classified into non-heat-treatable (work strain hardening) alloy and heat-treatable (age precipitation hardening) alloy capable of an increase in the strength due to the cold processing and the heat treatment, respectively.

Conventionally, it was difficult to laser welding Al and its alloys because the thermal conductivity of Al is about four times higher than that of Fe or steel and the laser reflectivity of Al is also higher. The penetration depths and surface bead widths of laser welds are compared among various alloys in Fig. 8.7 [7]. Easier melting and deeper penetration weld beads can be obtained in Al alloys with thermal conductivity reduced and a higher content of Mg, Zn, and Li, which are high volatile elements. A deep weld bead can be produced by the formation of a deep keyhole.

In Al alloys, porosity is easily formed in laser weld beads, as examples of CO_2 laser weld beads of A5182 alloy is shown in Fig. 8.8 [8]. At low speeds of 5–15 mm/s, porosity is absent or drastically reduced because almost all bubbles generated from

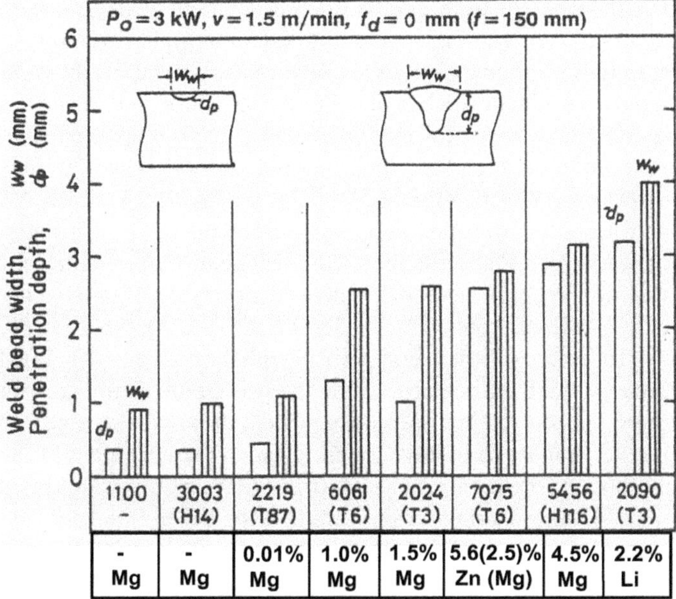

Fig. 8.7 Comparison of penetration depths and surface bead widths of CW CO_2 laser weld beads of various aluminum alloys

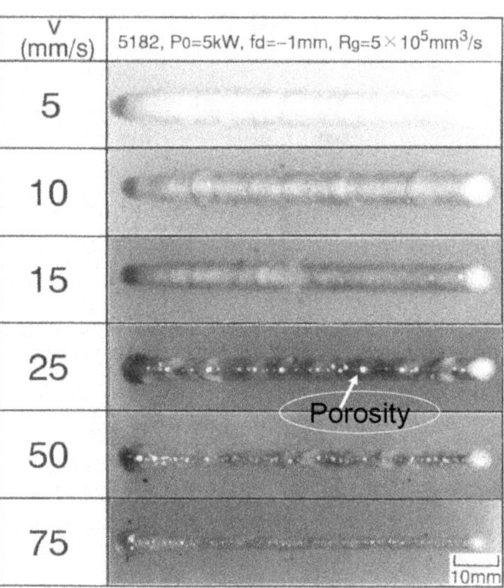

Fig. 8.8 Effect of welding speed on porosity formation tendency in CO_2 laser weld beads of A5182 alloy

a keyhole tip move back and upper to escape from the molten pool surface. At higher speeds, the size of pores or porosity is smaller due to the formation of smaller bubbles. It was possible to prevent porosity in the fiber laser weld bead of 3 mm depth at the power of 10 kW and extremely high speeds of more than 10 m/min. The porosity formation tendency of such easily melted aluminum alloys as a function of welding speed is different from that of steels and stainless steels since porosity is always present at lower welding speeds and absent at higher welding speeds (e.g., 5 m/min). In addition, although porosity is mostly easily formed at the welding speeds of 25 mm/s, pulse modulation is beneficial to the reduction in porosity, as the results of welds in A5182 alloy have been exhibited in Fig. 5.8. It is apparent that porosity is extremely reduced under the proper pulse modulation of about 70% in comparison with the CW laser welding conditions.

Mechanical properties of welds were evaluated by the tensile test, and some alloys were also assessed by the observation of the tensile testing specimens through the X-ray transmission imaging method. The tensile test results of laser welds of various aluminum alloys such as A5052-O, A5083P-O, A5182P-O, A6061P-T6, A6N01S-T5, and A7N01S-T5 are shown in comparison with the base alloys in Fig. 8.9 [9, 10]. In A5XXX alloys, the fracture occurred on the way of the strain-stress curve of the base alloy during tensile testing, and consequently, the tensile strengths of the welds are slightly lower than those of the base alloy and the strain values of the welds are considerably shorter. One example of X-ray transmission observation results of laser-welded A5182P-O alloy during tensile testing is shown in Fig. 8.10 [9]. The pores in the weld metal are deformed and elongated during tensile testing. The fractures in the tensile test specimens occurred from one large pore near the specimen surface or the neighboring large pore. The decreases in the strength and the strain of the welds are understood by considering the effect of pores or porosity on the reduction in the

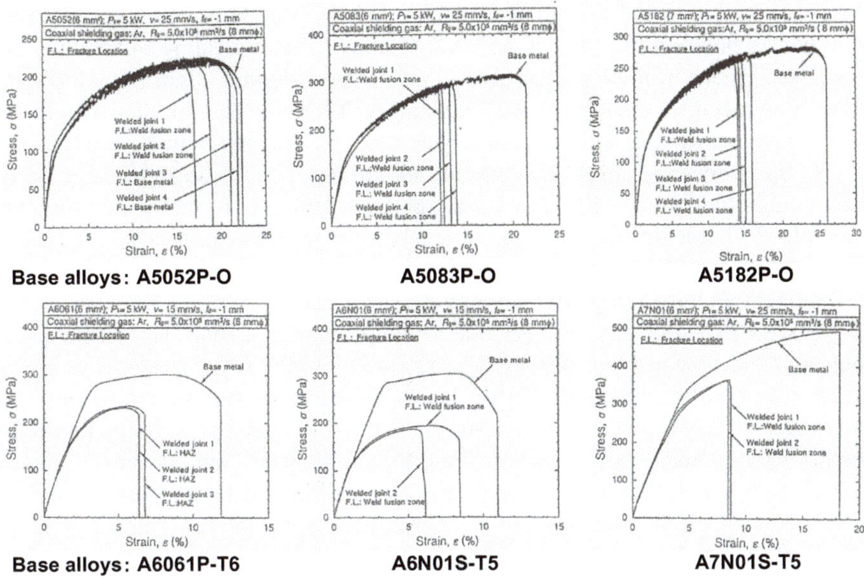

Fig. 8.9 Tensile test results of CO_2 laser welds and base metals of various aluminum alloys A5052-O, A5083P-O, A5182P-O, A6061P-T6, A6N01S-T5, and A7N01S-T5

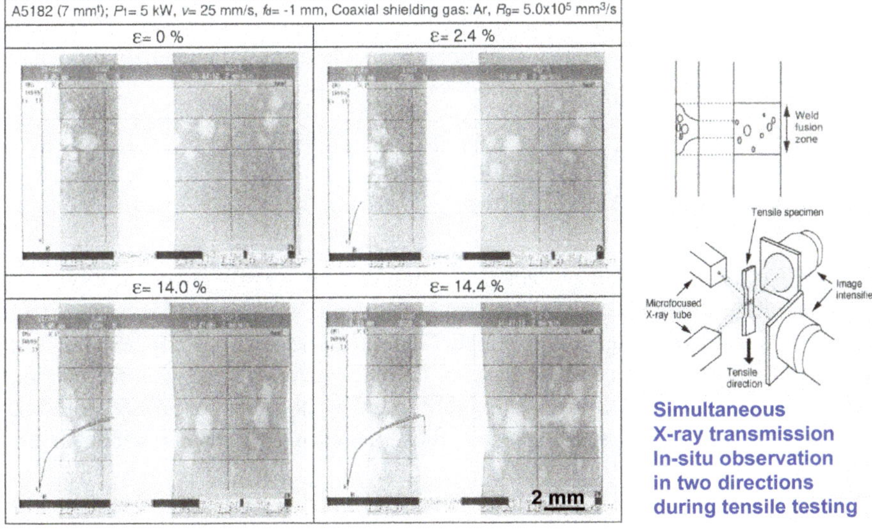

Fig. 8.10 X-ray transmission observation results and stress-strain curves of laser weld in A5182P-O alloy during tensile testing

cross-sectional area and the stress concentration near the large pores and between neighboring pores in the case of A5XXX-O alloy (after annealing treatment). The tensile specimens of laser-welded joints without porosity are almost equal to those of the base metal. In the case of A6XXX and A7XXX alloys, both the strength and the strain values of laser welds are largely decreased compared with those of the base alloys. These decreases of A6XXX welds are attributed to the disappearance of age precipitation hardening, difficult natural aging hardening at room temperature, and little solution hardening due to a low content of alloying elements such as Mg and Cu, although the strength of weld meta and HAZ of A7N01 alloy is slightly recovered by room temperature aging after welding. Besides, laser welds of A2024 and A7075 alloys become harder and stronger due to natural aging at room temperature with elapsing time after welding. Generally, in aluminum alloys, the hardness of a weld fusion zone is softer than that of HAZ, because the microsegregation always occurs and evaporation of Mg takes place, resulting in the shift from the optimum hardening conditions. It is noted that the welds of aluminum alloys should be weaker than the base alloy although the welds of steels are stronger than the base metal.

Currently, good surface appearances of laser weld beads can be formed in Al alloys by using ARM fiber laser with two cores of single and ring modes. It is confirmed that proper beam modes can suppress spattering and produce good welds with beautiful surface appearances, as already shown in Figs. 6.16 and 6.17 in Sect. 6.6.

In making battery cases, spot welding with pulsed YAG laser has been employed in terms of low heat inputs. Since general Al alloys are extremely sensitive to solidification cracking, only commercially available pure Al or A3003 Al alloy can be used for battery cases. Currently, CW fiber or disk laser of high beam quality can be employed at high welding speeds and low heat inputs.

8.3 Laser Welding of Copper

Copper (Cu) is mostly used as electrical materials because of high electrical conductivity and high thermal conductivity and is employed for electric car batteries, etc. Pure Cu is named “tough pitch copper” (electrolytic refining copper) including oxygen (O) made of electrolytic copper, highly conductive “oxygen-free copper” without oxygen (O), and so on. Cu alloys are called brass, phosphor bronze, aluminum bronze, etc., and used for water pipes, food processing machines, wires for laser brazing, and so on.

Laser welding of pure Cu is extremely difficult because of high thermal conductivity and high light reflectance. Welding of pure Cu is feasible by using high-power fiber or disk lasers, but there is a problem that large spatters are occasionally generated. Therefore, to suppress spattering and to satisfactorily produce weld beads in laser welding of Cu, investigations and studies are intensively performed by using high peak-power single-mode fiber lasers, ARM fiber lasers, blue lasers, and green lasers. The weld beads obtained with blue laser, general fiber laser and hybrid laser of fiber, and blue lasers have already been shown in Figs. 3.22 and 3.23. Thus, welds

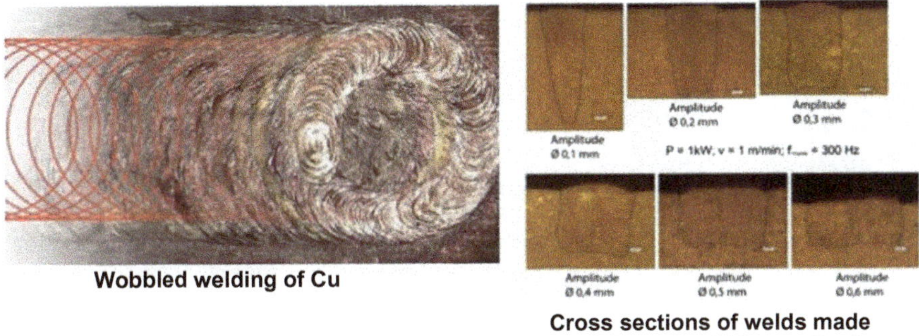

Fig. 8.11 Wobbled welding of Cu with single-mode fiber laser beam, and effect of wobble amplitude (width) on weld bead geometries

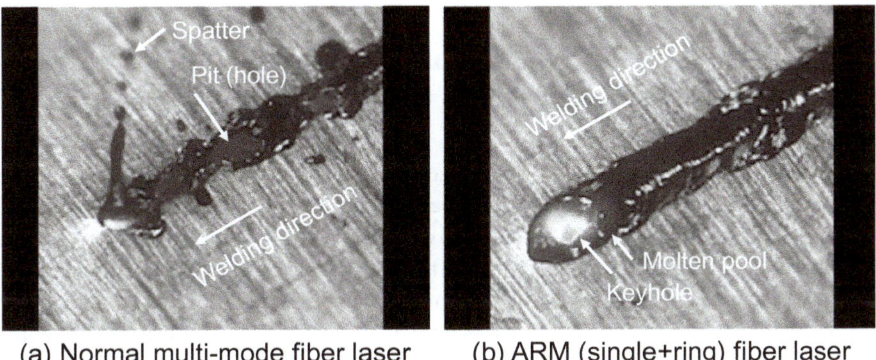

Fig. 8.12 Comparison of welding situation of Cu plate with normal fiber laser and ARM laser

produced with a single-mode fiber laser, and normal fiber laser and ARM laser are exhibited in Figs. 8.11 and 8.12 [11, 12]. Good weld beads of various penetration depths and bead widths can be produced by rotating (wobbling) a single-mode fiber laser beam under the proper rotating speeds and widths. An ARM fiber laser can weld a Cu plate more stable than a normal fiber laser to produce a good weld bead. It is reported that welding with a blue laser or a green laser is also more stable than that with a normal fiber or disk laser. Moreover, hybrid welding using a blue laser combined with a fiber laser is possible to produce a stable good weld bead. The above lasers and hybrid lasers are applied for welding Cu hairpins, many thin Cu sheets, Cu plates, etc., in industries, as shown in Fig. 8.13 [13, 14].

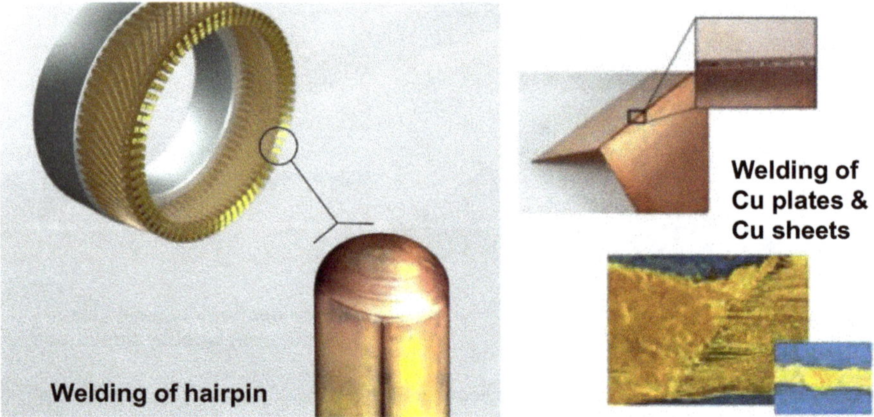

Fig. 8.13 Laser welding examples of Cu hairpins, Cu plates, and many thin Cu sheets

8.4 Laser Welding of Magnesium Alloys, Titanium, and Ni-Based Super Alloys

Magnesium (Mg) and its alloys are characterized by extremely low specific gravity of about 1.8 corresponding to 1/4 smaller value than steel, high specific strength (strength-to-weight ratio), and ultralightweight among practical metal materials. Therefore, Mg alloys are used for aircraft, cars, parts of electronic equipment, and so on. Their drawbacks are low corrosion resistance and low heat resistance, but in recent years, the corrosion resistance is enhanced by reducing impurities concentration due to the improvement in refining and smelting technology. Lately, Mg alloys have been used for cases of laptop computers, cameras, and mobile phones owing to good heat dissipation and electromagnetic shieldability. Mg alloys are made by wrought, extension or stretching processing, casting, diecasting, thixomolding, and so on.

It may be said that the laser welding of Mg alloys should be difficult, but this is judged to be wrong because a molten pool and a keyhole are easily formed. Figure 8.14 shows the welding conditions for the formation of good or bad full-penetration welds and partial penetration welds in AZ31 alloy [15]. Wrought Mg alloys are stably welded with lasers under the proper welding conditions, as examples of sound laser weld beads are shown in Fig. 8.15 [15]. A remarkable decrease in the tensile strength of laser-welded joints is not detected in wrought alloys. The strength of the weld bead made at higher welding speeds is slightly higher due to finer grains than that made at lower welding speeds. On the other hand, it is noted that porosity should be easily formed in diecasting, thixomolding, and a microsegregated part of casting, as shown in Figs. 8.16 and 8.17 [16]. The surfaces of these welds are poor, and the strengths of the welded joints are low. In Mg cast alloys, porosity is necessarily formed in the laser weld beads in the central part of specimens or parts, but porosity may be reduced in the sheath because of the reduced microsegregation of

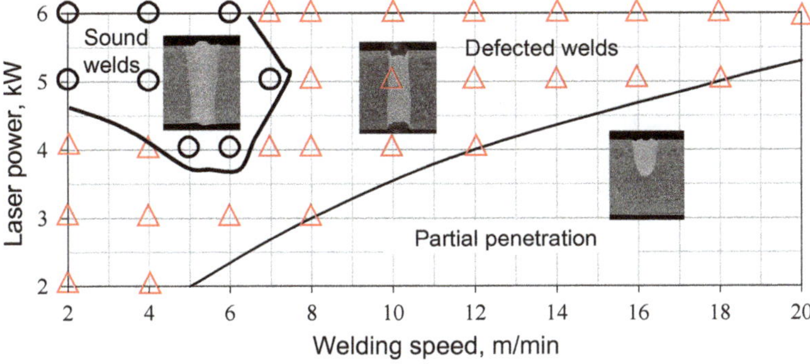

Fig. 8.14 Welding conditions of laser power and welding speed for formation of good or bad full-penetration welds and partial penetration welds in AZ31 alloy

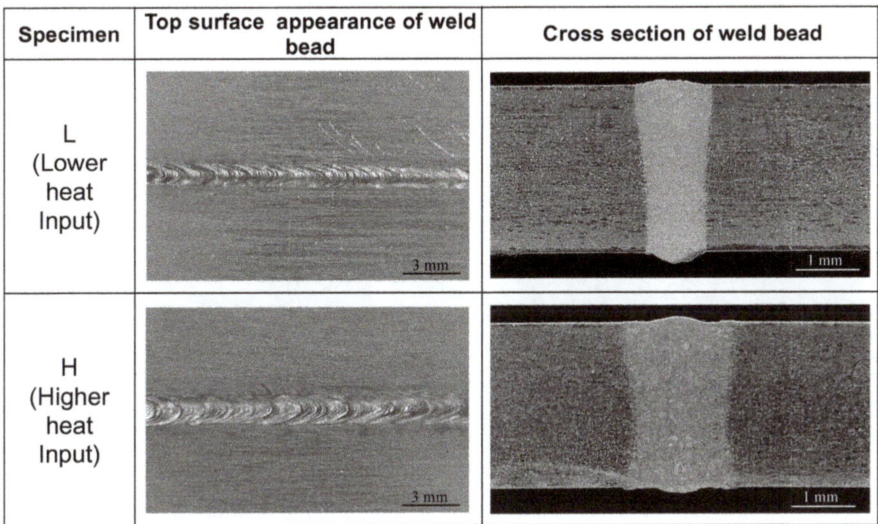

Fig. 8.15 Examples of sound laser weld beads in AZ31 alloy

impurities. Especially, in die-cast alloy, prevention of porosity is extremely difficult because of the existence of small-sized porosity origin. Laser welding of two passes or the use of an inserted rolled plate may be recommended to reduce porosity.

Ti and its alloys have good properties of the specific gravity of 4.5 corresponding to 40% lighter than steel, high strength, high ductility, high specific strength, and so on. Besides, since the strength is kept up to high temperatures of 723 K (450 °C), and Ti has a superior corrosion resistance and excellent biocompatibility, Ti and its alloys are used for aircraft, jet engines, gas turbines, sports goods, eyeglasses frames, medical equipment, and so on.

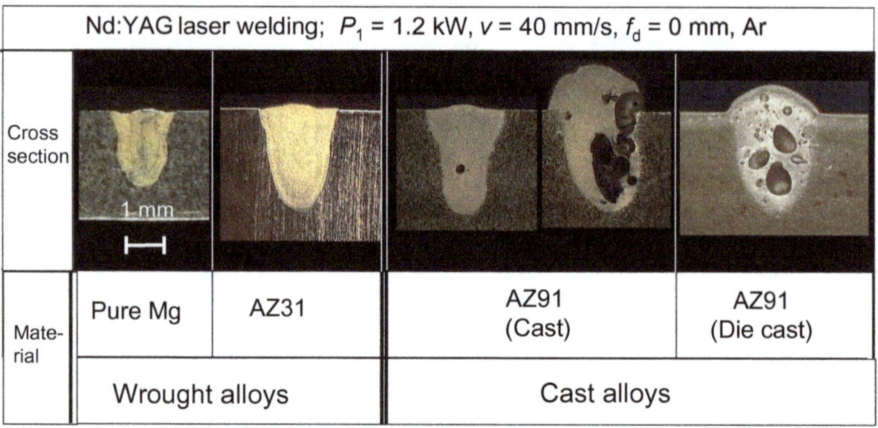

Fig. 8.16 Comparison of Nd:YAG laser welds of wrought and cast alloys, showing different formation tendency of porosity in respective alloys

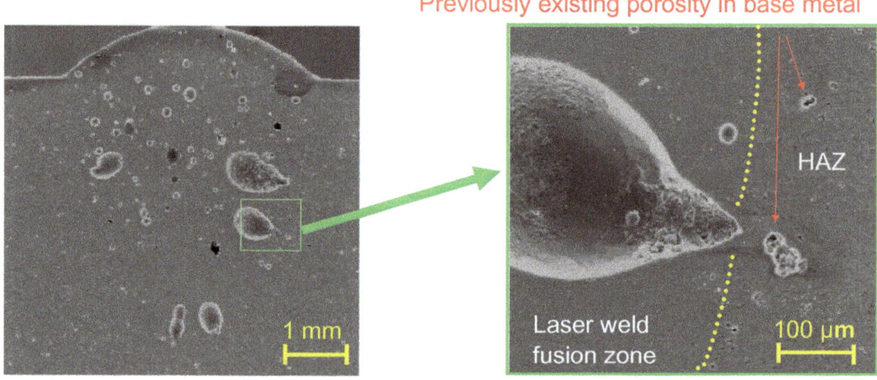

Fig. 8.17 SEM photos of Nd:YAG laser welds of die-cast alloy, showing characteristic porosity grown from small porosity in fusion boundary near HAZ

The drawbacks are in the following: (1) Ti and its alloys are brittle due to oxidation, nitriding, or carbonizing at high temperatures; (2) Ti and its alloys are brittle due to hydrogen absorption at more than 573 K (300 °C); (3) drilling or cutting of Ti and its alloys is difficult; and (4) Ti and its alloys are expensive. The microstructures of Ti alloys are classified into three types of alpha (α), beta (β), and α + β. Typical titanium and its alloy are α type of pure Ti of high corrosion resistance, and α + β type of Ti–6Al–4 V alloy of superior strength and toughness.

In laser welding of Ti and its alloys, the welds show acicular microstructure due to rapid cooling and are harder and stronger than base alloys, resulting in no problems. Nevertheless, examples of spatters and underfilled weld beads made in Ti plate with a CW laser at 10 kW and 50 mm/s are shown in Fig. 8.18 [17]. Spatters and porosity

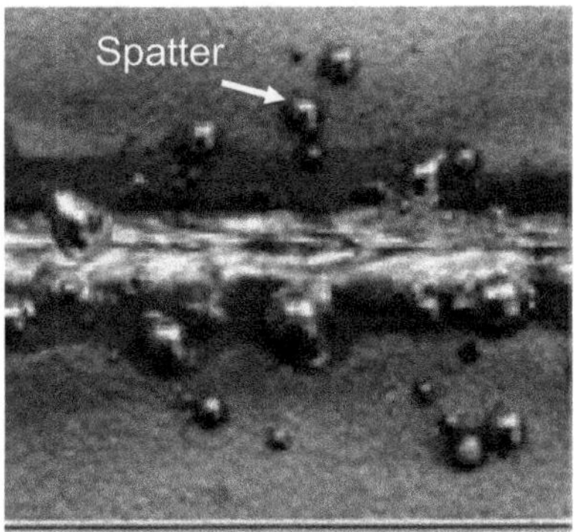

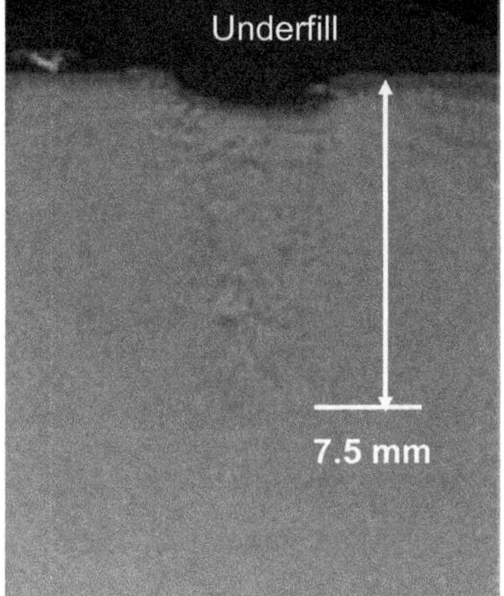

Fig. 8.18 Example of spatters and underfilled weld bead made in Ti plate with CW fiber laser at 10 kW and 50 mm/s

formation tendency in Ti seam welds made with a normal pulsed YAG laser and a pulse-shaped controlled YAG laser were compared, and the results are already shown in Figs. 4.9 and 4.10. It is apparent that a laser beam is easily absorbed and spattering is likely to occur during welding. It is therefore important to select the proper laser welding conditions, for example, a defocused distance, or a modified pulse shape. Pulse-shaped control of laser power is required to produce sound welds without spatters and porosity in spot welding with a pulsed laser.

Ni-base alloys are classified into Ni–Cu system, Ni–Al system, Ni–Fe system, Ni–Cr system, etc. Ni-base alloys are used for chemical plants, jet engines, etc., because they possess superior heat resistance and high corrosion resistance. Most Ni-base alloys are susceptible to hot cracking such as solidification cracking in weld fusion zones and liquation cracking in the waist part of HAZ near the fusion boundary in a wine cup type of a fusion zone. Especially, spot welds made with a pulsed YAG laser are extremely sensitive to solidification cracking along the grain boundaries. Hot cracking may be reduced under the proper welding conditions. Figure 8.19 shows cross-sectional and surface microstructures of CW YAG laser welds made in CM247LC single crystal at the power of 1 kW, the welding speed of 1 mm/s, and the

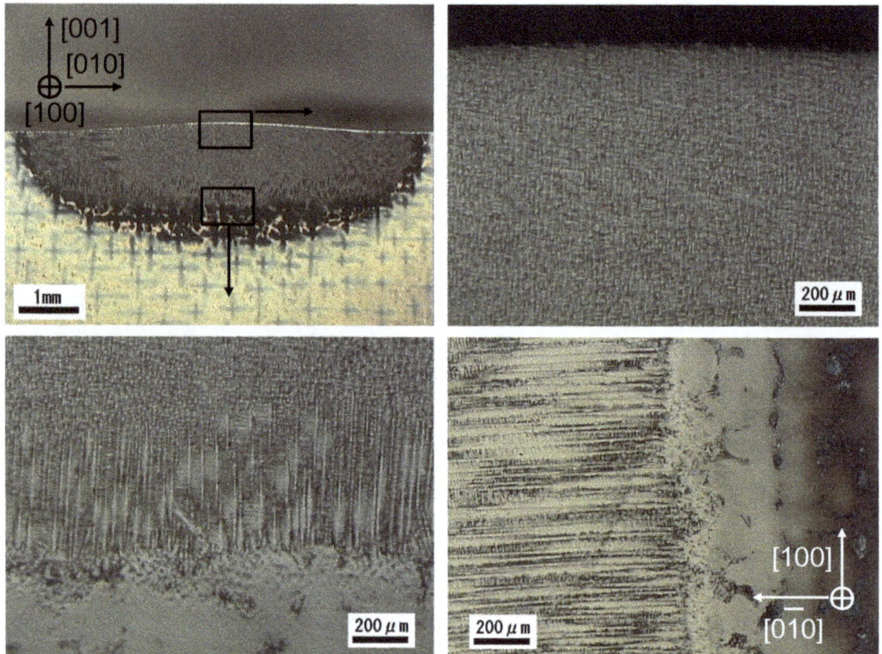

Fig. 8.19 Cross-sectional and surface microstructures of CW YAG laser welds made in CM247LC single crystal at 1 kW and 1 mm/s in Ar gas of 30 L/min, showing <100> solidification growth direction of cellular dendrites

defocused distance of 7.5 mm in Ar gas of 30 *L*/min, where the welding direction is parallel to [100] direction of the single crystal [18]. Most cellular dendrites grow in the <100> direction, but some equiaxed crystals are formed probably due to melt flows and microsegregated inclusions in the cellular dendritic boundaries of the base metal. In the case of single crystals, hot cracks (solidification cracks) are absent in shallow weld fusion zones probably due to the absence of long grain boundaries.

8.5 Laser Welding or Brazing of Ceramics

Extremely hard ceramics are produced by combining with some metals and oxygen (O), nitrogen (N), or carbon (C). Ceramics such as alumina (Al_2O_3), zirconia (ZrO_2), silicon-nitride (Si_3N_4), aluminum-nitride (AlN), and silicon-carbide (SiC) are called new ceramics or fine ceramics because of excellent properties such as high heat resistance, high corrosion resistance, high wear resistance, high hardness, and lightweight. Nevertheless, machining of these ceramics is difficult.

Joining of ceramics with CW CO_2 laser was tried. It was reported that CO_2 laser welding of Al_2O_3 with low purity was possible. To prevent hot cracking, heating up

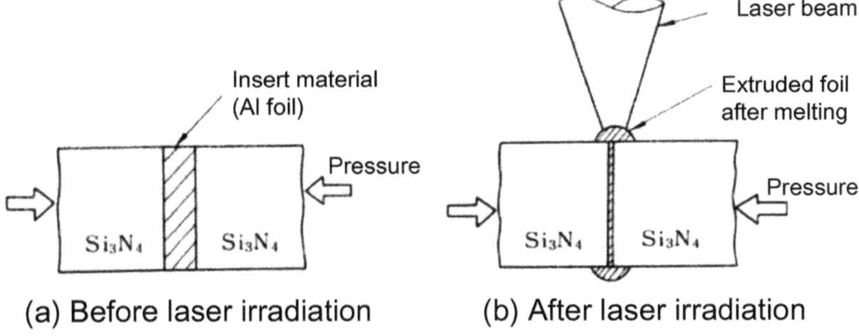

Fig. 8.20 Laser brazing of sublimable ceramics Si3N4 by using Al film

to high temperatures (generally, more than 1000 °C) is inevitable during welding, and slow cooling from high temperatures is required after welding. In joining of sublimable ceramics Si_3N_4, laser brazing was tried by using an Al foil, as shown in Fig. 8.20 [19, 20]. Laser brazing was reported to be possible to join Si_3N_4 plates.

8.6 Laser Joining of Plastics

Plastics are synthetic resins and are divided into "thermoplastic resins" softening by heating and "thermosetting resins" losing liquidity or fluidity. They are simply called "resins" including synthetic resins and natural resins or "polymers" owing to macromolecules. Plastics are used for various goods such as automobile or car parts, medical equipment, electronic components or parts because of super-lightweight, good corrosion resistance, and excellent electrical insulation.

Laser joining of thermosetting resins is impossible, but laser joining of thermoplastic resins such as polyimide (PA) or nylon is possible, as the laser lap jointing and samples of plastic sheets are shown in Fig. 8.21 and the joining process and mechanism are schematically indicated in Fig. 8.22 [13, 21, 22]. A transparent plastic plate is firmly set as an upper plate on the lower plate of plastic including carbon black capable of laser absorption (so-called, laser-absorbing plastic). A laser beam is transmitted through the upper transparent plastic plate and is absorbed near the upper surface of the lower absorbing plate. Consequently, the upper surface of the lower plate is melted, and thereby the bottom part of the upper plate adjacent to the melted zone near the lower plate surface is melted, resulting in laser lap joining of two plastic plates. Such joining with diode lasers is practically used for key cases, intake manifolds, oil server, engine oil sensor, and head, fog and tail lamps in cars, electric appliances, medical devices, and so on, as the application examples are shown in Fig. 8.23 [13, 14, 21–23]. It is recognized that the same kind of transparent and absorptive plastic plates can be easily joined. Figure 8.24 shows the effect of laser heat input on the joint strength of PA6 with glass fibers [22]. It is seen that the strong joints can be obtained under the proper conditions in the Range 2 of the figure. The

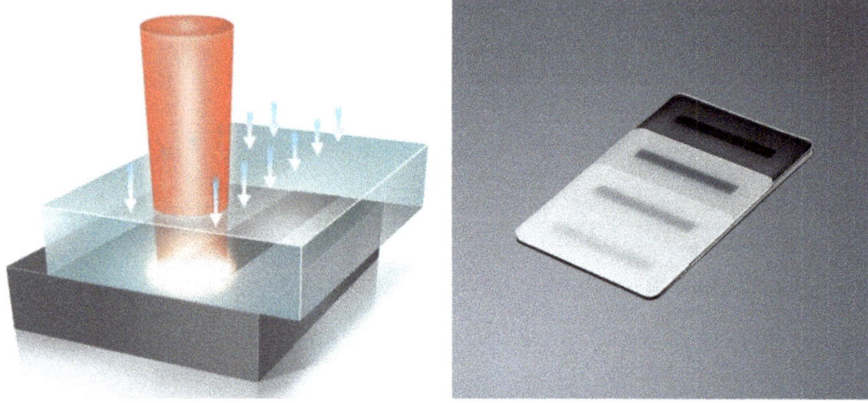

(a) Laser lap joining of plastic sheets (b) Several laser lap joints of plastic sheets

Fig. 8.21 Schematic laser lap joining process and examples of laser lap joints of thermoplastic resins

(1) A transparent plastic sheet is firmly set on the lower plate of laser absorbing plastic.

(2) A laser beam is irradiated on transparent plastic sheet, and laser absorbing plastic is soon heated and melted.

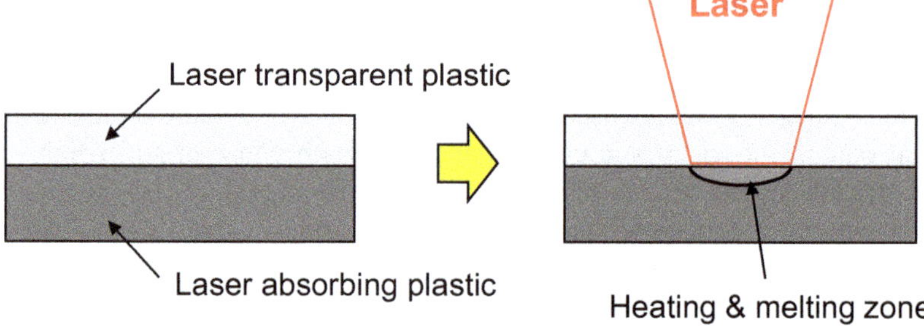

(3) A transparent plastic sheet is melted due to heat transfer from laser absorbing plastic.

(4) Natural cooling occurs due to heat transfer to the surroundings, and solidification joining occurs.

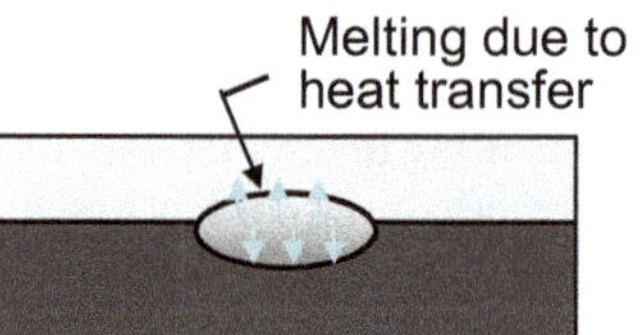

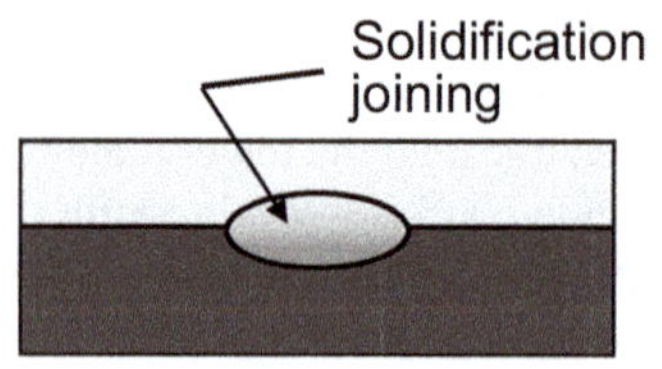

Fig. 8.22 Schematic illustration of laser lap joining process and mechanism of thermoplastic resins

Fig. 8.23 Various samples of laser lap joined thermoplastics

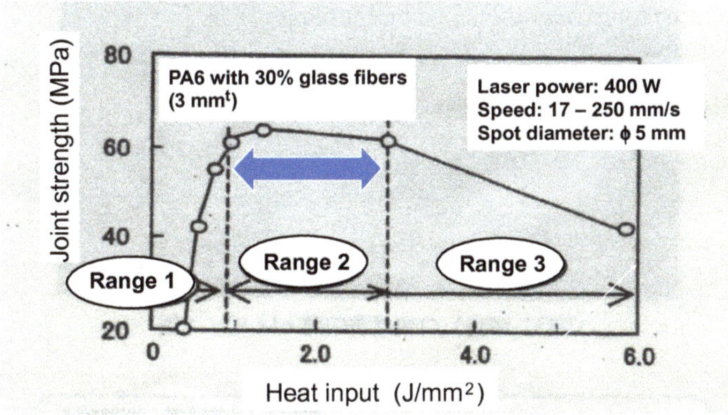

Fig. 8.24 Effect of laser heat input on tensile shear strength of lap joints of PA6 plastic sheets with glass fibers

joint strengths are lower because the bonding areas are narrower at lower heat inputs (Range 1), and since the formation tendency of porosity increases at higher heat inputs (Range 3). Besides, it is often said that dissimilar plastics have difficulty to be joined in comparison with similar plastics. In such a case, a slightly higher laser power (or a slower welding speed leading to higher heat input) may be recommended for lap joining of dissimilar plastics to form a small number of bubbles to enhance melt flows for mixing in the molten pool, which is similar to the mechanism of joining of metal and plastic plates as described in Sect. 9.6.

Moreover, thermoplastic resins including partially laser-absorbing substances are developed for butt joining [24]. Good joints of thermoplastic resins are produced under the proper laser joining conditions.

8.7 Laser Welding or Joining of Glass

Glasses have characteristics of uniformity, isotropy, and disorder. The glasses have an absorption edge due to electron transition between band gaps in the ultraviolent side, and an absorption edge due to lattice vibration in the infrared side. The light in the region between both edges is transmitted, and the glass is transparent.

Consequently, a CO_2 laser in the far-infrared region is absorbed in glasses and melting near the plate surface is possible to join glasses [25]. Moreover, a femtosecond laser is focused to achieve extremely high-power density near the focal point, and thus this laser can produce lap joints of two glass sheets or plates, as an example of lap joint of glass has already been shown in Fig. 3.24 and schematic lap joining of glass is indicated in Fig. 8.25 [26]. Laser joining of glasses is possible by making melting zones near the lapped joint planes [25].

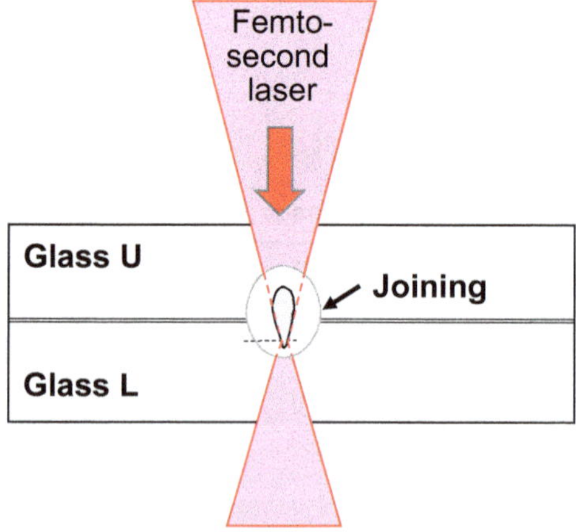

Fig. 8.25 Schematic lap joints of two glass sheets or plates with femtosecond laser

References

1. Katayama's Laboratory-3, JWRI, Osaka University
2. Pic A, Munera DD, Vierstraete R, Pinard F (2006) In: Proceedings of the 67th laser materials processing conference, JLPS, pp 49–62
3. Katayama S, Yoshida D, Matsunawa A (2004) J High Temper Soc HTSJ, 30(2): 93–99 (in Japanese)
4. Katayama S, Fujimoto T, Matsunawa A (1985) Trans JWRI 14(1):123–138
5. Katayama S, Matsunawa A (1984) Proc ICALEO '84. LIA 44:60–67
6. Katayama S, Matsunawa A (1985) Proc ICALEO '85. LIA 44:19–25
7. Katayama S, Lundin CD (1991) J Light Metal Welding JLWA 29(8):349–360 (in Japanese)
8. Katayama S, Kojima K, Matsunawa A (1998) J Light Metal Welding, JLWA 36(1):3–21 (in Japanese)
9. Katayama S (2006) J Light Metal Welding, JLWA 44(8):333–343 (in Japanese)
10. Katayama S, Kojima K, Matsunawa A (1999) J Light Metal Welding, JLWA 37(2):47–58 (in Japanese)
11. Grupp M, Reinermann N (2017) Proc. LiM 2017, WLT
12. Odajima T (2018) Personal Communication Data; Mitsui Bussan Electronics Ltd.
13. Nakamura T (2018) The latest laser welding system of TRUMPF. In: Proceedings of the 89th laser materials processing conference, Osaka, Japan, vol 89, pp 55–60 (in Japanese)
14. Takeda S (2019) Personal communication, Laserline. Web (https://www.laserline.com/)14
15. Wahba M, Mizutani M, Kawahito Y, Katayama S (2010) Sci Technol Weld Join 15(7):556–566
16. Morita S, Katayama S, Matsunawa A, Hino M High Energy Density Research Committee of Japan Welding Society (JWS), EBW-495–01 (in Japanese)
17. Nakamura H, Kawahito Y, Nishimoto K, Katayama S (2015) J Laser Appl LIA, 27(3):032012-1-10
18. Sakamoto M, Katayama S (2003) In: Proceedings of international symposium on novel materials processing by advanced electromagnetic energy sources, ASM
19. Miyazaki T, Miyazawa H, Murakawa M, Yoshioka S (1991) Laser processing technology. Sangyo-Tosho (Industry-Library), Ltd., p 18 (in Japanese)
20. Maruo H, Miyamoto I et al (1989) Reprint Japan Weld Soc Ann Meet JWS 44:132–133
21. LEISTER Web Homepage. (https://wwe.leister.com/jp/)
22. Nakamura H, Terada M (2003) J Japan Weld Soc JWS 72(3):189–192 (in Japanese)
23. Brockmann R, Nakamura T (2020) Personal communication, TRUMPF Laser- and Systemtechnik, GmbH. Web (https://www.trumpf.com/)
24. Orient Chemical Industries, Ltd. Web Homepage (www.Ltw.jp/index.php)
25. Miyamoto I (2013) Laser welding of glass. In Katayama S (ed) 'Handbook of laser welding technologies'. Woodhead Publishing, pp 301–331
26. Katayama S (2019) Very easy book of laser processing. The Nikkan Kogyo Shimbun, Ltd. (in Japanese)

Chapter 9
Laser Welding, Joining, or Brazing of Dissimilar Materials

9.1 Laser Welding of Steel and Cast Iron

Laser welding of a low alloy steel and a cast iron was investigated in place of fixing bolts with nuts. Figure 9.1 shows the examples of fiber laser welds of low alloy steel SCM 420 and spheroidal graphite cast iron FCD 600 of 30 mm thickness [1]. Laser irradiation location was controlled by tracking under the sensing apparatus system shown already in Fig. 7.2 [2]. Consequently, in the case of the shift of 0.07 and 0.15 mm to the cast iron from the butt–joint interface, quenching cracking occurred along with the Fe_3C cementite in the HAZ of cast iron and across weld beads. On the other hand, in the case of the butt–joint interface and 0.07 mm shift to the low alloy steel, quenching cracking could be prevented, but solidification cracking occurred in the lower part of deep weld fusion zones. In the latter case, the plates of 8 mm in thickness were used, and thus sound welds without cracking could be produced, as shown in Fig. 9.2 [1]. Moreover, when the laser was irradiated along the butt–joint interface under the defocused conditions, weld beads without solidification cracks could be produced with narrow widths near the bottom [1]. Moreover, a wire with a high content of Ni was used for welding of cast iron, and it was consequently confirmed that sound laser-welded joints without cracks could be also produced in joining low alloy steel and cast iron.

Laser welding of dissimilar steels of low alloy steels and stainless steels or ferritic and austenitic stainless steels is relatively readily feasible, and the compositions and microstructures of laser weld fusion zones can be anticipated from the mixing ratios of respective steels and Schaeffler's diagram.

S. Katayama, *Fundamentals and Details of Laser Welding*,
Topics in Mining, Metallurgy and Materials Engineering,
https://doi.org/10.1007/978-981-15-7933-2_9

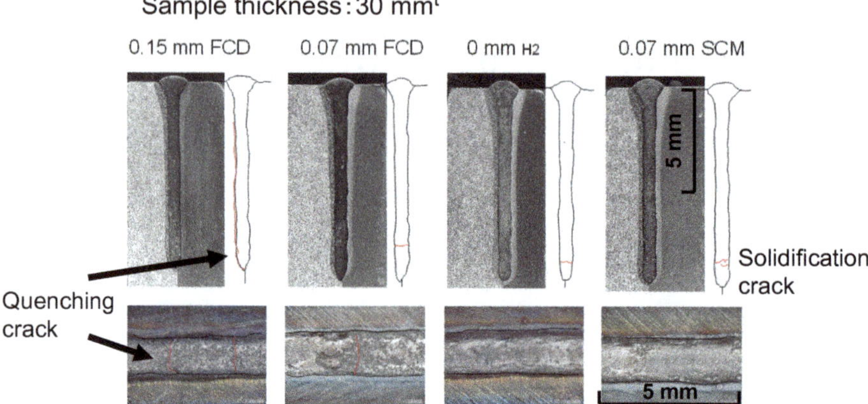

Fig. 9.1 Examples of fiber laser welds of low alloy steel SCM 420 and spheroidal graphite cast iron FCD 600 of 30 mm thickness made at different laser irradiation positions, showing quenching cracks and solidification cracks in respective welds

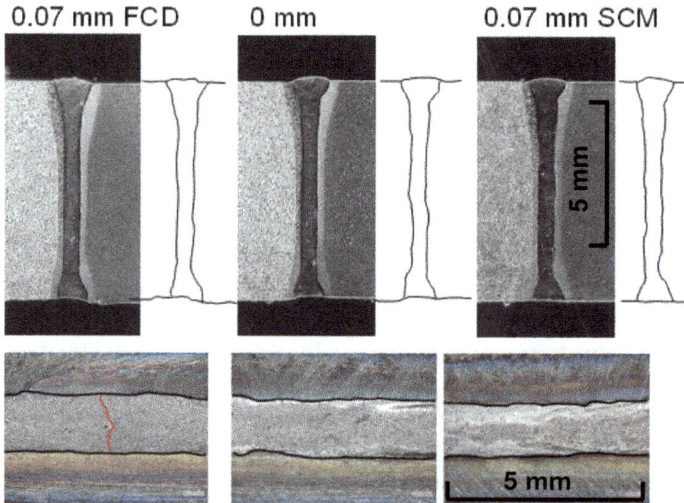

Fig. 9.2 Examples of fiber laser welds of low alloy steel SCM 420 and spheroidal graphite cast iron FCD 600 of 8 mm thickness produced at different laser irradiation positions, showing quenching cracks in welds at laser beam shift of 0.07 mm to FCD side and no cracks in welds at laser beam shifts of 0 and 0.07 mm to SCM steel side

9.2 Laser Welding of Steel and Aluminum Alloy

Laser dissimilar welding of steel of the highest usage and aluminum alloy of lightweight is receiving the most attention. Thus, welding of low carbon steel and Al–Mg–Si alloy A6xxx with a diode laser or a YAG laser was first tried. An example

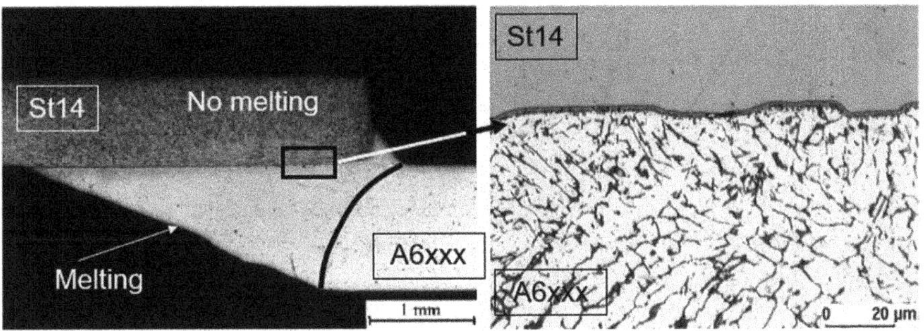

Fig. 9.3 Example of cross section and microstructure near joining interface of CW YAG laser weld in joint of steel St14 and Al alloy A6xxx, showing joining due to melting of A6xxx alloy lower sheet

of cross section and microstructure near the joining interface of CW YAG laser weld in joint of steel St14 and Al alloy A6xxx is shown in Fig. 9.3 [3, 4]. The laser was shot on the upper steel sheet, and then the sheet was heated up to high temperatures. The heat could be conducted to the lower Al alloy sheet, and then Al alloy sheet was melted because of its low melting temperature to weld the upper steel sheet. Flux was required to remove an oxide film on the Al alloy sheet when the sheet could be melted by the heat conduction type. A sound weld bead without defects was produced, and the intermetallic compound layer was about 2 μm thin. It was reported from the tensile test that the fracture occurred in the HAZ of A6XXX alloy and a strong welded joint could be made.

Moreover, laser lap welding of steel and aluminum alloy was investigated under various conditions. A steel or stainless steel sheet is set as the upper plate, and an Al alloy sheet is put as the lower plate because of the suppression of wide-area formation of intermetallic compounds. The tensile shear test results of the lap-welded joints of 0.8-mm-thick Fe and 1-mm-thick A1050 sheets are shown in Fig. 9.4 [5]. The high tensile load or strength of welded joints could be obtained under the proper conditions. At low and proper heat inputs, the Fe component flowed into the Al sheet due to the formation of a keyhole and the melt flows from the keyhole tip near the bottom of the molten pool. The intermetallic compounds of FeAl, Fe_2Al_5, etc., were formed along the boundary between penetrated Fe and melted Al, as Fe–Al binary phase diagram is shown in Fig. 9.5 [6]. It has been reported that intermetallic compounds of $FeAl_2$, Fe_2Al_5, and $FeAl_3$ are brittle [7]. The fracture of the specimen occurred along the intermetallic compounds and/or the neighboring Al weld fusion zone. Thus, the load or strength depended upon the depth and width of Fe weld fusion zone penetrated and flowed into the Al sheet. At low heat inputs, the tensile shear loads (or strengths) were low because the fracturing easily occurred because of shallow penetration depths in Al sheet and narrow weld bead widths at the interface. On the other hand, at high heat inputs, intermetallic compounds were largely and widely formed in the weld fusion zones, and the cracks in welding and fracturing in the test occurred in such intermetallic compounds, resulting in low tensile shear load or strength. Consequently, in laser welding of steel and Al alloy, it is important to

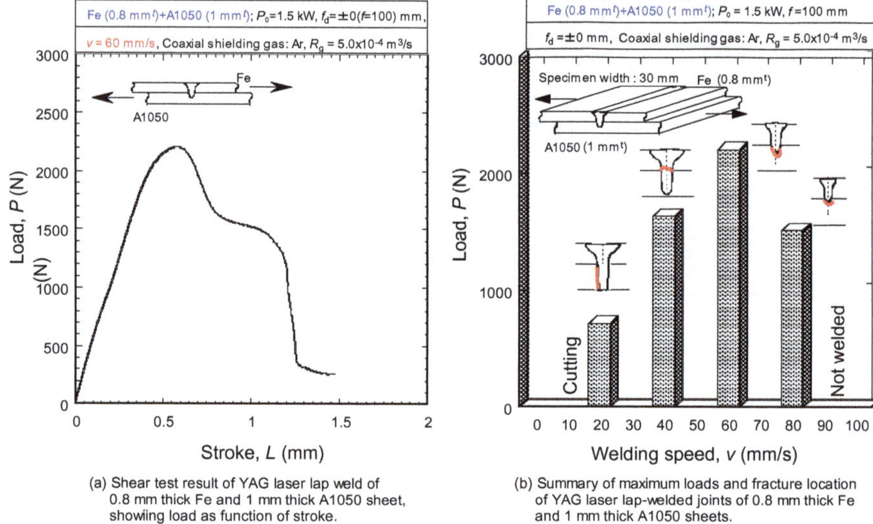

(a) Shear test result of YAG laser lap weld of 0.8 mm thick Fe and 1 mm thick A1050 sheet, showiing load as function of stroke.

(b) Summary of maximum loads and fracture location of YAG laser lap-welded joints of 0.8 mm thick Fe and 1 mm thick A1050 sheets.

Fig. 9.4 Stress–strain curve of one test piece and tensile shear test results of lap joints of 0.8-mm-thick Fe and 1-mm-thick A1050 sheets welded with CW YAG laser at 1.5 kW, showing effect of welding speed on tensile shear load or strength

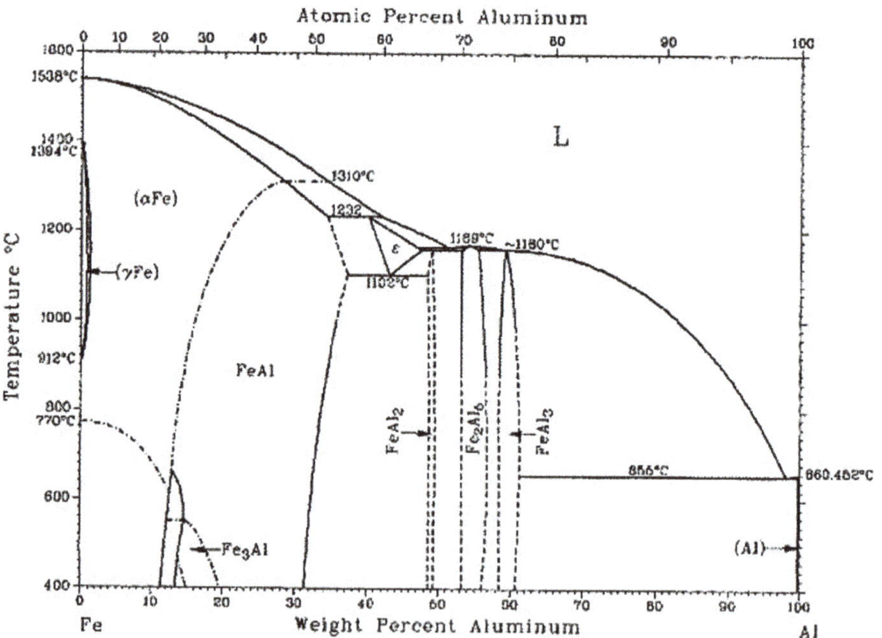

Fig. 9.5 Fe–Al binary phase diagram, showing formation of various intermetallic compounds at high temperatures

select the proper welding conditions. The strong welded joints, in which the fracture occurred in Fe sheets, could be produced by increasing the joining area due to, for example, the production of three weld beads under the proper conditions, as shown in Fig. 9.6 [8]. This is a remarkably simple laser dissimilar welding method, but the steel penetrated in Al alloy is limited to be about 0.2–0.4 mm. Besides, butt and lap simultaneous welding of steel and Al alloy was developed to produce a wider joining area by one welding pass. The special joint was developed, as schematically shown in Fig. 9.7 [9]. The cross sections of a weld bead of SPCC and Al alloy A5052 are exhibited in Fig. 9.8 [9]. Butt and lap joint of higher strength was produced. This was applied to make a pipe of stainless steel and Al alloy.

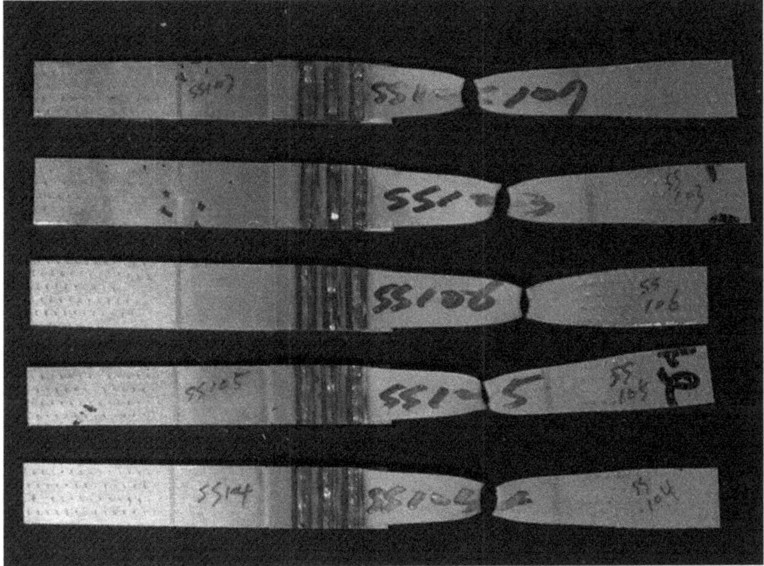

Fig. 9.6 Tensile shear test samples of three laser lap weld beads made under proper conditions, showing fracture in SPCC base specimen

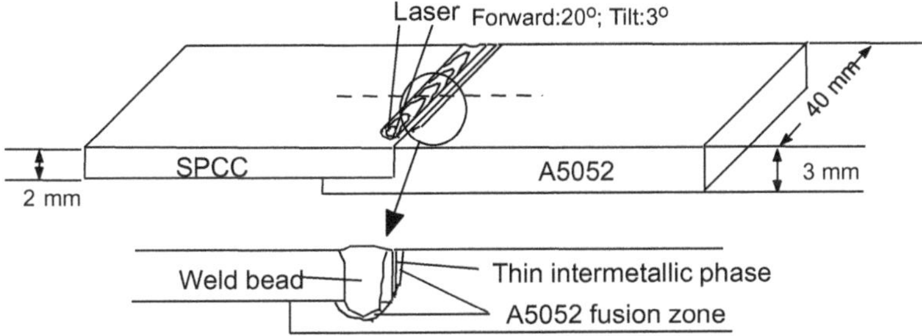

Fig. 9.7 Schematic special sample for simultaneous laser butt and lap welding of steel and Al alloy

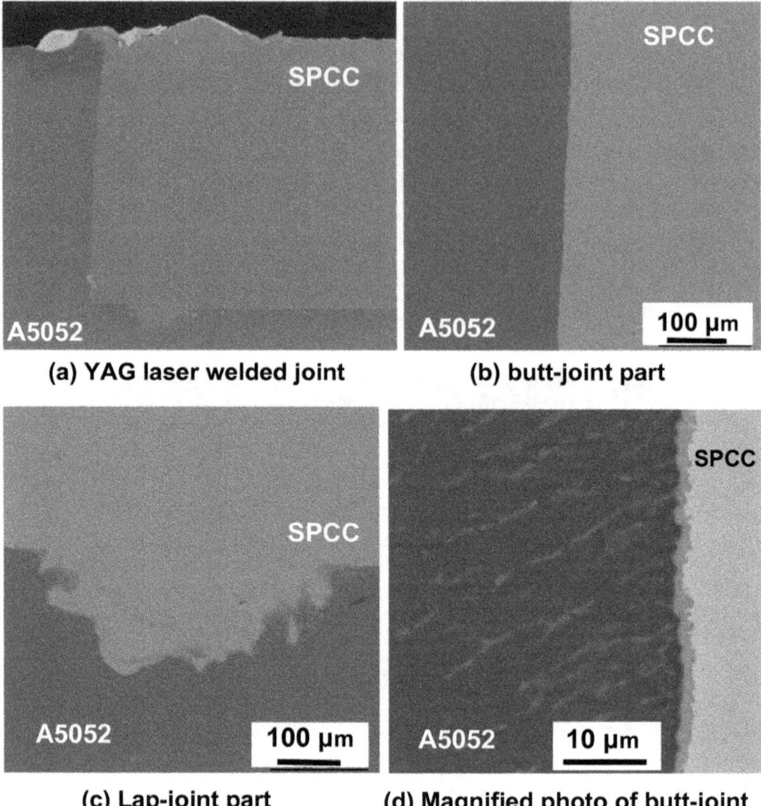

Fig. 9.8 Cross sections of laser weld bead of simultaneous butt and lap joint in special specimens of SPCC and A5052 sheets

Laser brazing of steel and Al alloy was also investigated by using Al–12%Si wire. Strong joints could be produced under the conditions of proper laser powers and heat inputs. At low heat inputs, the joining of Al alloy and Al wire was difficult. At high heat inputs, thick brittle intermetallic compounds were formed between steel plate and a brazed fusion zone of Al alloy wire. The use of flux for the removal of oxide film covering Al alloy plate should be required.

9.3 Laser Welding of Steel and Copper

Concerning laser welding of steel and copper (Cu), welding of austenitic stainless steel and pure Cu with CW CO_2 laser or CW or pulsed YAG laser was performed by changing the irradiation location in various shielding gases. Sound strong welds could be produced by irradiating most of a laser beam on the stainless steel plate, because Cu

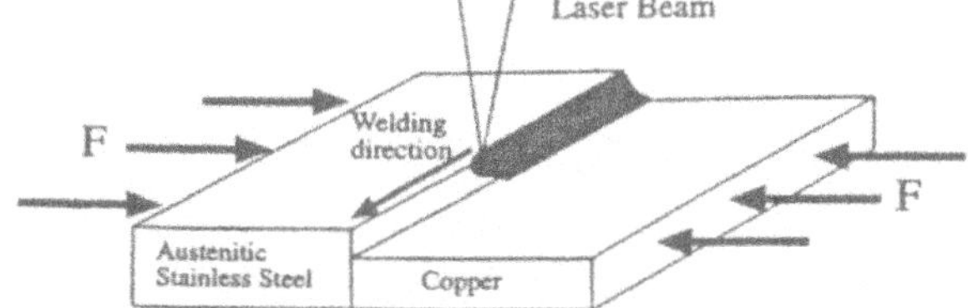

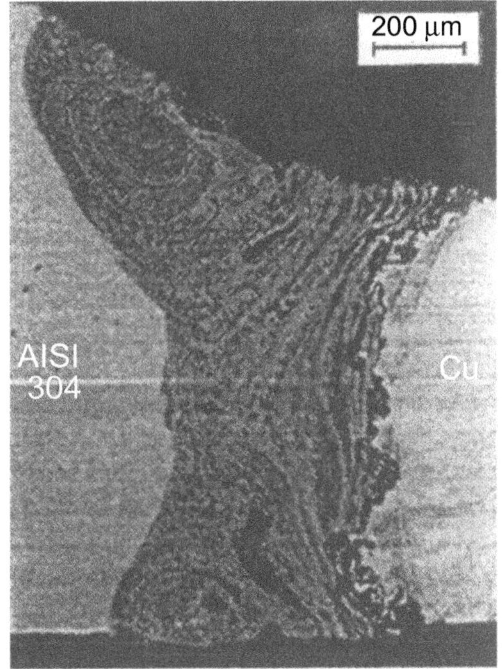

Fig. 9.9 Schematic of dissimilar stainless steel and copper sheets and cross section of laser weld bead in Type 304 and Cu butt-joint sheets

was highly reflective and greatly heat-conductive, as shown in Fig. 9.9 [10]. Normally, the fracture of the tensile test occurs in the HAZ of Cu, where work-hardening effect disappears due to the thermal history. Intermetallic compounds are not formed in the weld fusion zone between steel and Cu, as Fe–Cu phase diagram is shown in Fig. 9.10 [6], and thus, relatively strong weld joints can be produced. Nevertheless, when both metals are melted similarly, the primary solidification of austenite phase occurs accompanied with Cu microsegregation over the wide temperature range. Consequently, solidification cracking or liquid metal embrittlement cracking may occur, which should be noted.

Laser welding was applied to make a pipe of dissimilar Type 304 and 70Cu–30Zn brass. Lap welding was successfully performed by irradiating a laser on Type 304 upper pipe under the proper control of the penetration depth. Besides, two passes may produce a better weld bead by remelting cracks for repairing.

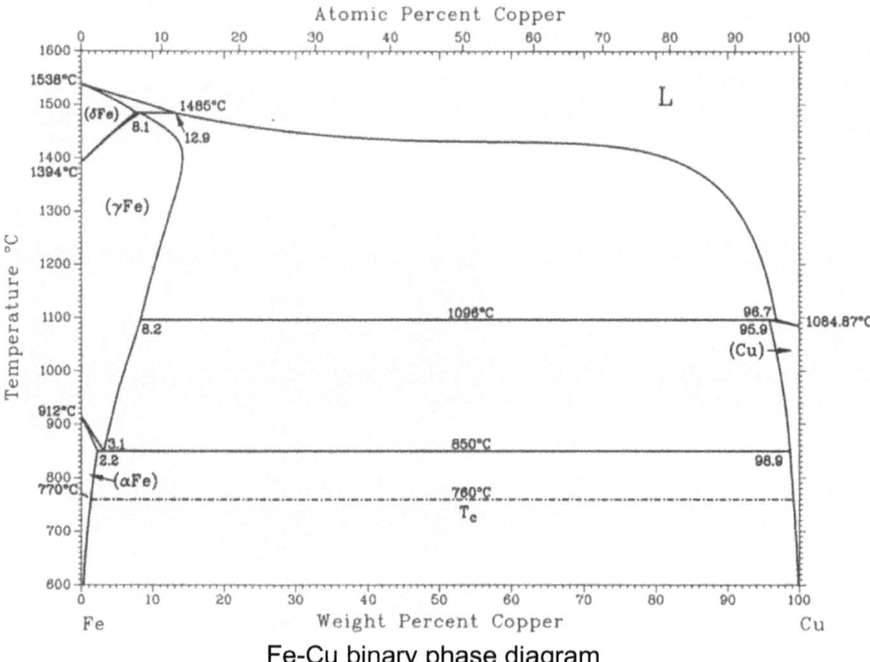

Fig. 9.10 Fe–Cu binary phase diagram, showing wide liquidus and solidus temperature range and no formation of intermetallic compounds

9.4 Laser Welding of Steel and Magnesium Alloy

There are no intermetallic compounds between iron (Fe) and magnesium (Mg) just similar to the formation phases between Fe and Cu. Therefore, there is no problem of the harmful effect of compounds on mechanical properties.

The results of the tensile test of laser-welded joints between steel SPCC and Mg alloy AZ31 and austenitic stainless steel Type 304 and AZ31 are shown in Fig. 9.11 [11]. Fairly high strengths (high stresses required for fracture) were obtained in the butt joints of dissimilar materials made with a CW YAG laser and a CW CO_2 laser. It is interesting to know that a laser beam should be irradiated on the steel plate to produce a normal weld fusion zone, and the heat from the steel weld molten pool can melt the Mg alloy plate adjacent to the butt–joint interface, as shown schematically in Fig. 9.12 [11]. Namely, joining is performed just like liquid phase diffusion bonding or welding between solid steel and melted Mg alloy. In fact, it seemed that Al in Mg alloy could easily diffuse into steel, helping to join steel and Mg alloy. Similar welding results should be obtained with diode, disk, and fiber lasers.

Laser lap welding of 3-mm-thick AZ31 Mg alloy plate and 1-mm-thick SPCC steel sheet with or without Zn-coated film was investigated by using a CW disk laser or a CW fiber laser. As a result, in the case of Mg alloy upper plate, severe

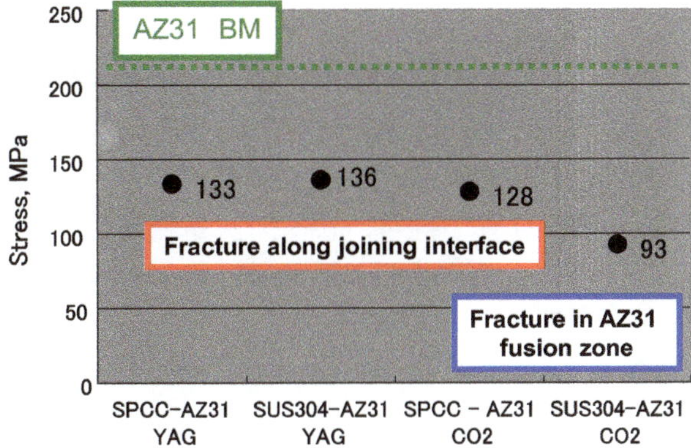

Fig. 9.11 Tensile strengths of butt joints between SPCC or Type 304 and AZ31 alloy welded with YAG laser or CO_2 laser

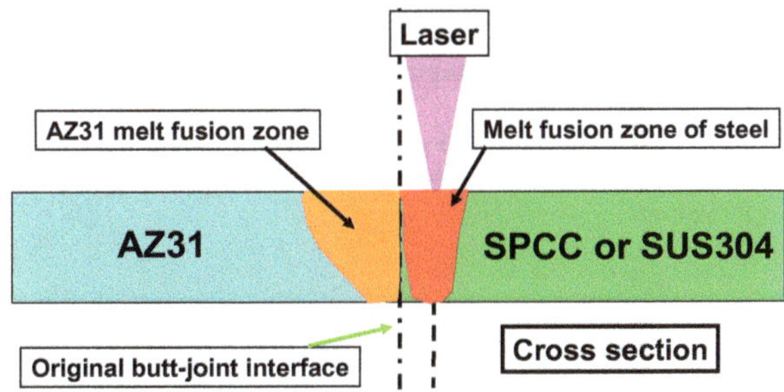

Fig. 9.12 Recommended laser butt-joining process of steel or austenitic stainless steel and Mg alloy AZ31 sheets

spattering leading to underfilling occurred to produce a very narrow joining area, and the strength of the joint was low. On the other hand, in the case of Zn-coated SPCC upper sheet, a strong welded joint was obtained although bare SPCC upper sheet without Zn-coating film was not welded with AZ31 alloy lower plate, as shown in Fig. 9.13 [11]. Macro- and microstructure of the cross section of a laser-welded joint are exhibited in Fig. 9.14 [12]. Zn-coated steel is melted in a heat conduction mode, and the heat from the melted steel melts not only the bottom Zn-coating film but also low melting point Mg alloy plate beneath the melted steel. Then Mg–Zn eutectic layer is formed near the interface. Al in Mg alloy may diffuse into steel to support joining at the interface, probably resulting in the formation of Fe_3Al extremely thin film, and Fe particles are present in Mg alloy layer adjacent to the steel bottom.

From such joining results of dissimilar steel and Mg alloy, a laser beam should be used to melt high melting-point steel, and Mg alloy adjacent to the steel is melted by the heat from a steel molten pool. In the case of lap welding, Zn-coating film must be effective. Moreover, Al in Mg alloy should exert a beneficial effect on the joining of Mg alloy to steel.

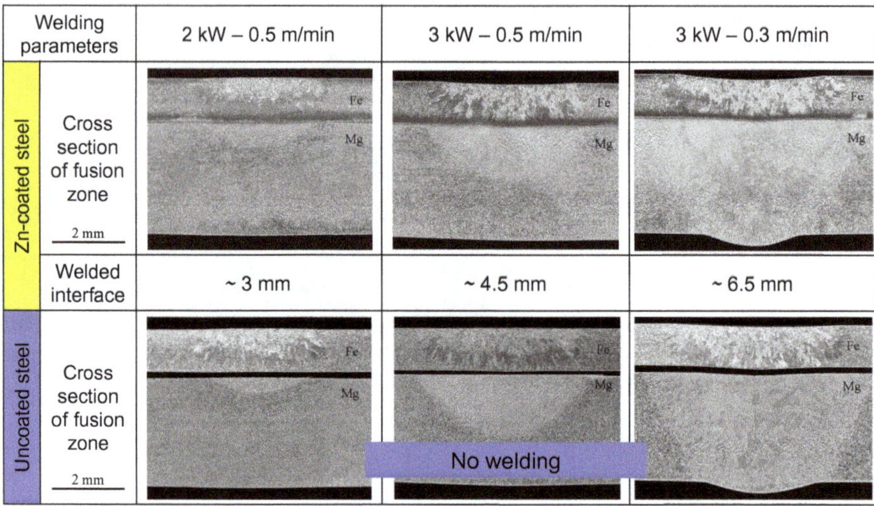

Fig. 9.13 Cross sections of lap joints of SPCC upper sheets with or without Zn-coating and AZ31 alloy lower plate, showing effect of Zn-coating on joining and no welding between bare steel and AZ31 alloy

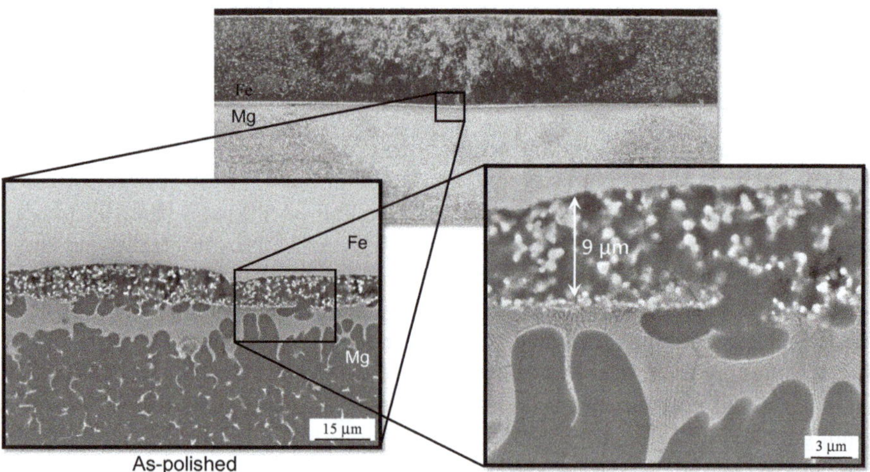

Fig. 9.14 Macro- and microstructure of cross section of strong laser lap-welded joint, showing formation of Mg–Zn eutectic phase near interface between Zn-coated steel and AZ31 alloy

9.5 Laser Welding of Copper and Aluminum Alloy

Recently, dissimilar lap- or butt-joint welding of pure Cu and pure Al has been investigated for electrodes (conductor: bus bar) of large-sized Li-ion batteries by using a pulsed YAG laser, a CW disk laser and a CW fiber laser, accompanied with

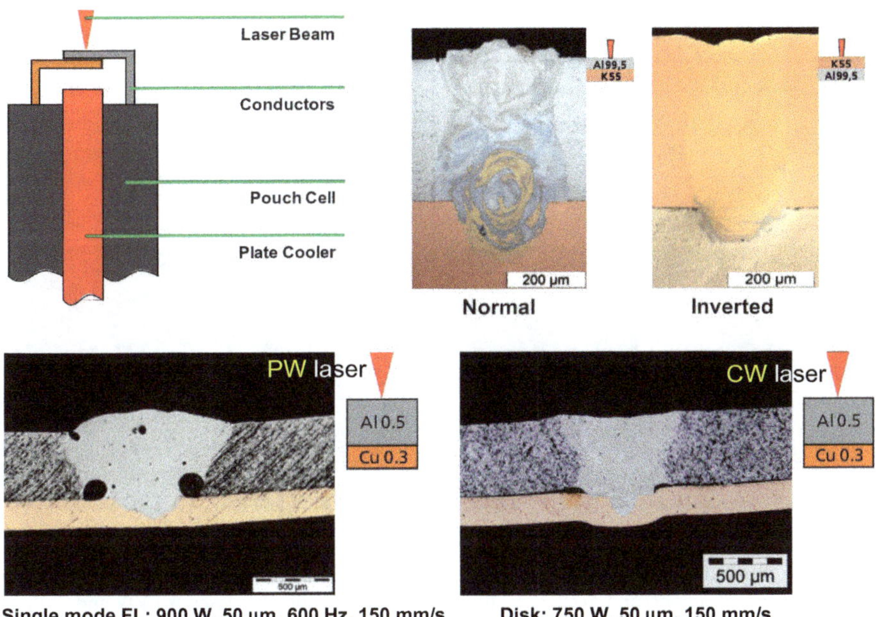

Fig. 9.15 Cross sections of lap weld joints of Cu (upper)–Al (lower) and Al (upper)–Cu (lower) sheets made with fiber or disk laser, showing formation of partial penetration weld

the development in hybrid cars and electric cars. Examples of lap welding of Cu (upper)—Al (lower) and Al (upper)—Cu (lower) sheets are shown in Fig. 9.15 [13]. A laser beam is often irradiated on the Al upper sheet because laser melting of Al is easier than that of Cu. But when the laser power is too high, spatters should occur, leading to underfilled welds, or brittle intermetallic compounds are widely formed, resulting in the formation of cracking and the reduction in the strength of joints. If Cu can be stably melted, the laser irradiation on Cu upper sheet should be recommended since the formation of intermetallic compounds may be reduced. It is important to select the appropriate welding conditions for the suppression of intermetallic compounds in laser welding of Cu to Al.

9.6 Laser Welding of Titanium and Aluminum Alloy

It is expected to take advantage of superior properties of strong and high corrosion resistant Ti and its alloys and lightweight Al and its alloy. Lap welding of Ti alloy (Ti6Al4V) and Al alloy (AlMg0.4Si1.2) was investigated by shooting CW YAG laser or diode laser on Ti alloy sheet [4]. Al alloy was melted by the heat from the laser-heated Ti alloy, and a strong joint was produced by joining a solid Ti alloy sheet and a molten Al alloy sheet so as to form a thin intermetallic compound film in the

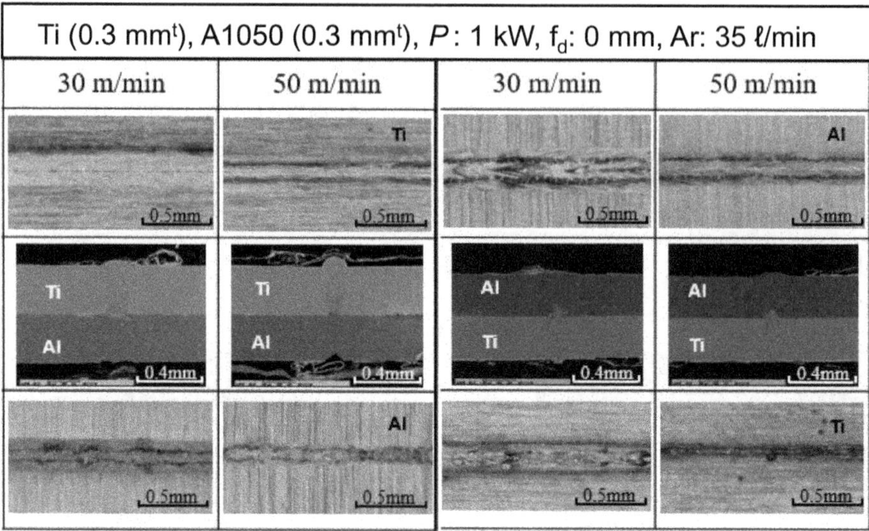

Fig. 9.16 Top and bottom surfaces and cross sections of lap weld beads made in Ti (upper) and Al (lower) and Al (upper) and Ti (lower) sheets with single-mode fiber laser at 1 kW and at 30 and 50 m/min

interface. The tensile test result showed that fracture occurred in the HAZ or base metal of Al alloy, and the strength of the joint was about 220 MPa [4].

Lap welding of pure Ti sheet of 0.2 mm thickness and pure Al sheet of 0.2 mm thickness was investigated by using a single-mode fiber laser at the high speeds of 5–50 m/min. The weld bead and intermetallic compounds were observed, and the joint loads (strengths) were evaluated. The top and bottom surfaces and cross sections of weld beads made in Al (upper) and Ti (lower) sheets at 1 kW and at 30 and 50 m/min are shown in Fig. 9.16 [14]. Sound welded joints without cracks could be produced at the high speeds of more than 30 m/min, although cracking occurred at the low welding speeds of less than 20 m/min probably due to the formation of large intermetallic compounds of Ti and Al. The tensile shear loads of lap welds made in dissimilar metals at various speeds are shown together with those of similar Al–Al or Ti–Ti in Fig. 9.17 [14]. The tensile shear loads of Al–Ti and Ti–Al sheets are between Ti–Ti and Al–Al, and the decrease in the loads is small from low speeds to high speeds although the loads of similar metal sheets decrease with an increase in the welding speed. In the case of similar metals welding, the decrease in the strength is attributed to the decrease in the weld bead width or joining area near the interface of both sheets. On the other hand, the tensile shear loads of dissimilar metals are higher than those of Al–Al sheets but lower than those of Ti–Ti sheets. In dissimilar joints, the flow-in of strong Ti elements in soft Al weld metal suggests a relationship of higher strengths. This result shows that relatively strong welded joints can be produced by lap welding of dissimilar Al and Ti sheets and Ti and Al sheets with a

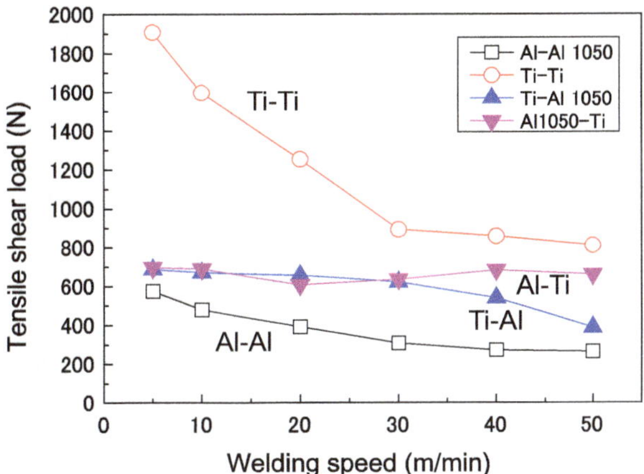

Fig. 9.17 Comparison of various tensile shear loads of lap welds made in dissimilar Ti & Al sheets and similar Ti & Ti and Al & Al sheets at various speeds of 5 to 50 m/min, showing stronger joints in Ti–Al sheets than Al–Al ones

single-mode fiber laser at high welding speeds by penetrating Ti element into an Al fusion zone and by suppressing the formation of intermetallic compounds.

9.7 Laser Joining of Metal to Plastic or CFRP

Metals and plastics are used in almost all industries such as cars, automobiles, railroad vehicles, airplanes, electrical machineries and electronic equipment, and joining or bonding of metals to plastics is important inevitable technology. Generally, dissimilar materials of metals and plastics are bonded with adhesive agents or mechanically fastened with screws, bolts, rivets, etc. However, epoxy and acrylic adhesives are used with organic solvents, which are volatile and harmful to the health of workers. Therefore, serious problems are that they are included into VOC (volatile organic compounds) regulation, and besides that it takes long time to perform adhesive bonding. On the other hand, the drawbacks of mechanical fastening are that the freedom of design is limited and that it needs another processing process and other parts.

Laser direct joining of metals and plastics is expected to overcome the above-described conventional problems and drawbacks [15, 16]. Figure 9.18 illustrates a schematic experimental method [16–18]. The plastics used for laser joining are thermoplastic resins but not thermosetting resings. Examples of laser direct joining of Type 304 stainless steel plate and PET plastic sheet are shown in Fig. 9.19 [16–18]. A transparent plastic sheet can be set as upper or lower plate. In the case of the use of a transparent plastic upper sheet, a diode, fiber, or disk laser is irradiated on the plastic sheet, part of transmitted laser is absorbed into the metal, and then the irradiated part of the metal is heated at high temperature but not melted. Thereby the plastic on the metal plate is melted by heat conduction from the heated metal plate and some parts of the melted plastic are decomposed to activate molten plastic and bubbles (which is

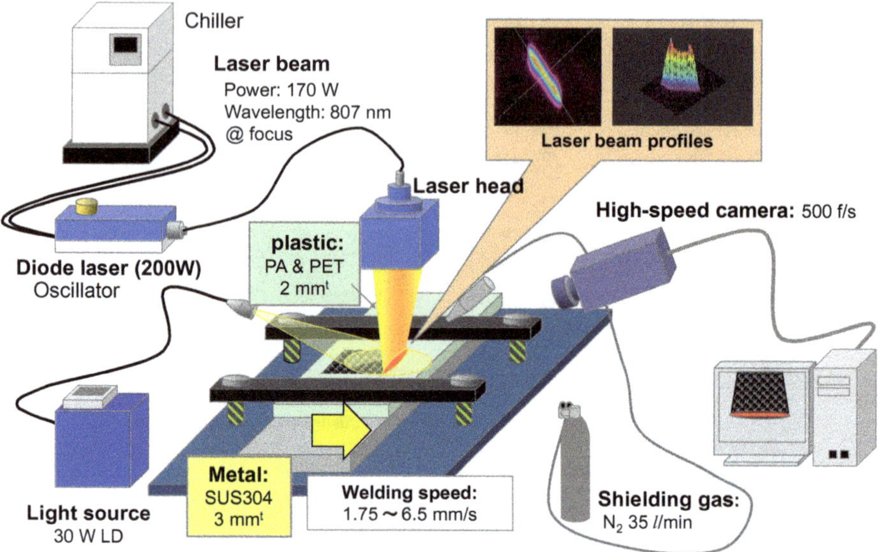

Fig. 9.18 Schematic experimental method and situation for laser joining of dissimilar metal and plastic plates

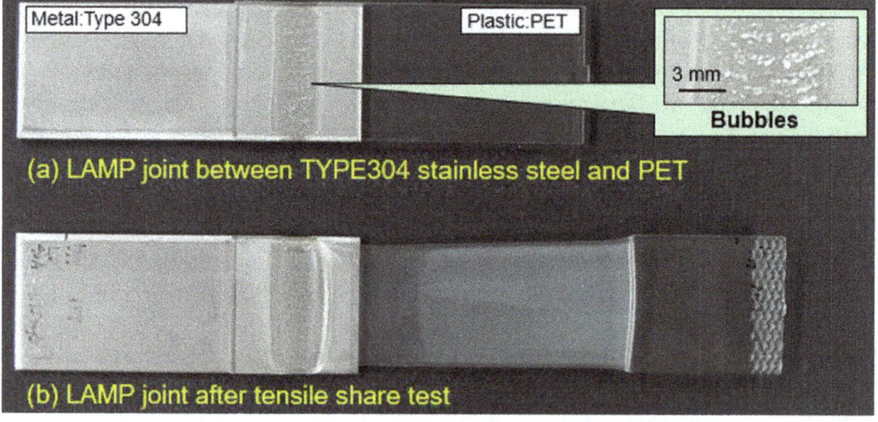

Fig. 9.19 Examples of diode laser direct joints of Type 304 stainless steel plate and PET plastic sheet before and after tensile testing

used in place of porosity or pores in plastics). High-pressured bubbles are generated to force the molten plastic to move to the metal surface, and it may be feasible to join activated metal and activated plastic heated at high temperatures. A joining or bonding area with small bubbles is observed. It is also confirmed that the joint is as strong as the plastic sheet because the plastic sheet is elongated and fractured in the tensile shear test. On the other hand, in the case of non-transparent plastics, the

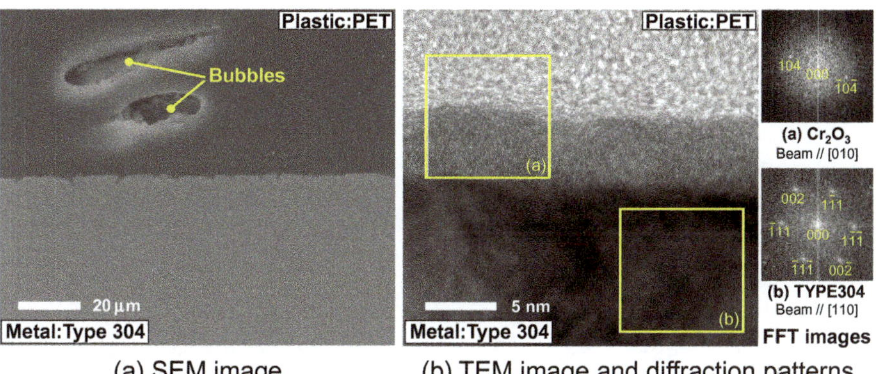

Fig. 9.20 SEM and TEM photographs of laser lap joint near interface between PET and Type 304 sheets, and diffraction patterns from metal and oxide film

metal sheet is set as an upper plate, and a laser beam may heat the metal sheet or produce a weld bead in the metal sheet. Consequently, the temperature of the metal bottom surface is raised to melt the beneath plastic sheet and sometime to make bubbles in the plastic melted zone. Thus, joining of dissimilar metal and plastic can be performed.

The cross sections of a lap joint of plastic and metal sheets were observed by scanning electron microscopy (SEM) and transmission electron microscopy (TEM) and analyzed by EDX, diffraction patterns of TEM, time of flight secondary ion mass spectrometry (TOF-SIMS), and Fourier transform infrared spectroscopy; attenuated total reflection (FT-IR-ATR). Figure 9.20 shows SEM and TEM photographs of a laser lap joint between PET and Type 304 sheets, and diffraction patterns from metal and oxide film [16–18]. It is understood that metal and plastic are firmly bonded with (Cr_2O_3) oxide film in atomic or molecular order.

Strong joints between metal and plastic sheets are attributed to three mechanisms: (1) anchor effect of rough metal surface covered with molten plastic; (2) chemical bonding with oxide film on the metal surface; and (3) van der Waals force due to nearing atoms and molecules, as schematically represented in Fig. 9.21 [16, 17]. Laser joining of metal and plastic sheets is feasible in any metals but is limited to engineering thermoplastics such as polyamide (PA), polyethylene terephthalate (PET), and polycarbonate (PC).

Laser joining of metal to plastic was also applied to dissimilar metals joining [19]. For example, PET (upper) to steel (lower) plate were joined and then aluminum (upper) to PET (lower) sheets were joined, leading to Fe–Al joining by using an inserted PET sheet. A strong dissimilar metals joint between steel and aluminum alloy was demonstrated [19].

CFRP has excellent properties for the applications in various fields. PA with fibers is non-transparent CFRP, and thus, a laser beam is irradiated on the surface of metal upper sheet and then the heat of the metal can melt the plastic of CFRP. It was

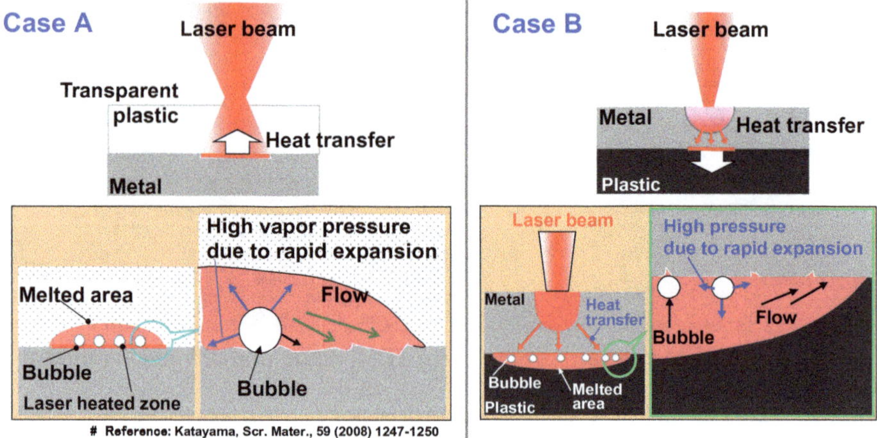

Fig. 9.21 Schematic illustration of laser lap joining processes of dissimilar metal and plastic sheets, showing melting and small bubbles formation in molten pool of plastic sheet adjacent to metal surface. (Case A: Laser irradiation on plastic sheet) (Case B: Laser irradiation on metal sheet)

revealed that joining of Type 304, aluminum alloy and Zn-coated steel to PA-based CFRP was feasible by melting the PA plastic in CFRP sheet adjacent to the metal sheet. The tensile shear test results of joints between Type 304 of 30 mm width and PA-based CFRP of 20 mm width are shown in Fig. 9.22 [20]. A strong joint of the metal and CFRP made under the proper conditions was confirmed. Bubbles were formed in CFRP, and thus, the fracture in the tensile shear test occurred through the

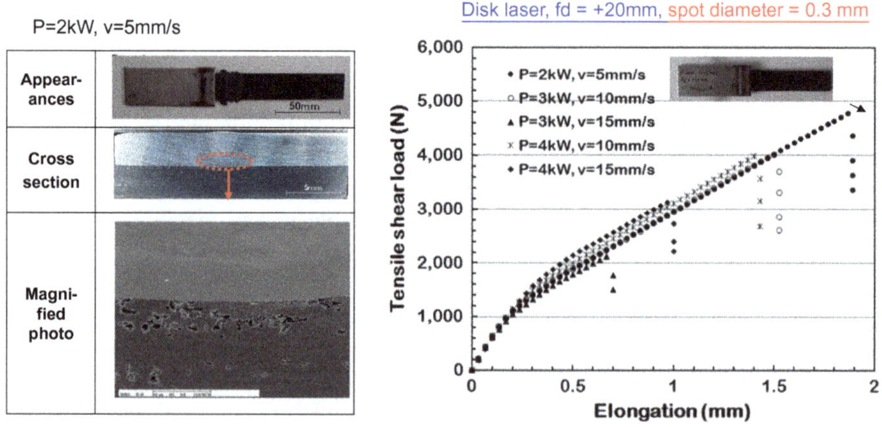

Fig. 9.22 Example of laser lap joint after test and tensile shear test results of lap joints made between dissimilar Type 304 of 30 mm width and PA-based CFRP of 20 mm width under conditions of various laser powers and traveling speeds

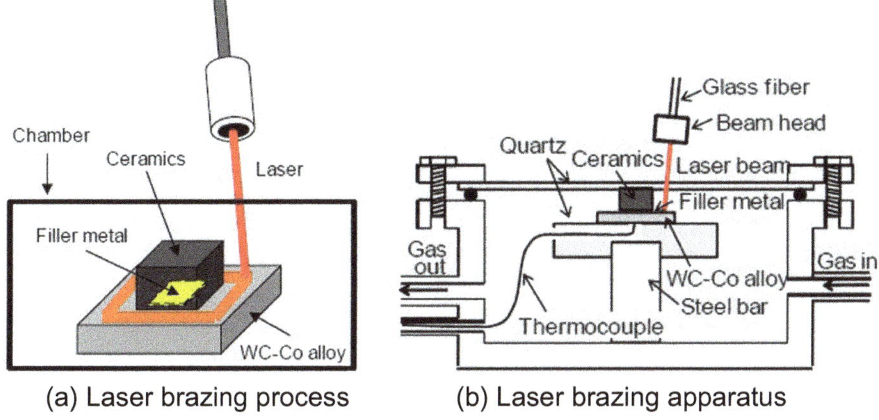

Fig. 9.23 Schematic illustration of laser brazing process and apparatus for dissimilar metal and ceramic sheets with filler metal foil

bubbles. The formation of a strong lap joint is attributed to the bonding of oxide film of the metal and thermoplastic (PA) in CFRP and the formation of a wide joint area.

9.8 Laser Joining or Brazing of Metal to Ceramic

To take a good use of superior properties of ceramics and metals, laser brazing with Ag–Cu–Ti filler metals was developed [21]. Experimental apparatuses are schematically illustrated in Fig. 9.23 [21]. h-BN, 99 mass% SiC, and sialon (90 mass% Si_3N_4) (of 5 mm × 5 mm × 3.5 mm) and WC-Co alloy (of 10 mm × 10 mm × 2 mm) were used as ceramic and metal, respectively. About 70–72% Ag-about 28% Cu-0 to 2.8% Ti filler wire was also utilized, and the effect of Ti content on oxidation and joining strength was investigated. Low vacuum and subsequent Ar gas flow, or a proper amount of Ar shielding gas were required to prevent the harmful effect of oxidation of Ti. Consequently, strong lap joints between ceramics and metals could be produced under the conditions of very low oxygen content.

References

1. Vondrous P, Katayama S, Dunovsky J (2010) Proceedings of the 74th laser materials processing conference, JLPS, pp 89–98
2. SERVO-ROBOT (3D Robot Vision Systems) Pamphlet, (Web Homepage; https://servo-robot.com)
3. Schubert E, Zerner I, Sepold G (1998) Proceedings of ICALEO '98, LIA, vol 85, Session G, pp 111–120
4. Wagner F, Zerner I, Sepold G (2001) Proc ICALEO 2001, LIA, 1301 (CD ROM)

5. Katayama S, Usui R, Matsunawa A (1998) Proceeding of 5th international conference on trends in welding research, Georgia, pp 467–472
6. Fe–Al Binary Phase Diagram; Fe–Cu Binary Phase Diagram; Web site
7. Yasuyama M, Ogawa K, Taka T (1996) J Japan Weld Soc JWS 14(2):314–320 (in Japanese)
8. Joo S-M, Kim Y-P, Bank H-S, Katayama S, Hwang W-S (2004) Key engineering materials. Adv Nondestruct Eval Part 3(270–273):2389–2394
9. Katayama S, Mizutani M (2003) Proc ICALEO 2003, LIA, Section E, (1401) (CD-ROM)
10. Lampa C, Powell J, Magnusson C (1997) Proc ICALEO '97, LIA 83(2):171–180
11. Katayama S, Morita M, Matsunawa A (2002) In: Naka M (ed) Proc. DIS '02 Designing of interfacial structures in advanced materials and their joints, Osaka, pp 747–750
12. Wahba M, Katayama S (2012) Mater Des 35:701–706
13. Gedicke J, Mehlmann B, Olowinsky A, Gillner A (2010) Proc ICALEO 2010, LIA, 103:844–849 (CD-ROM)
14. Lee S, Nakamura H, Kawahito Y, Katayama S (2012) Proc LMP 2012, Washington DC, #12-67, pp 1–5 (On-line proceeding of JLPS website)
15. Katayama S, Kawahito Y, Tange A, Kubota S (2006) Online Proc LAMP 2006, JLPS, #6–7
16. Katayama S (2010) The review of laser engineering. Laser Soc Japan 38(8):594–602 (in Japanese)
17. Katayama S, Kawahito Y (2008) Scripta Mater 59(12):1247–1250
18. Kawahito Y, Katayama S (2010) Proc 29th ICALEO 2010, LIA, 103:1469–1473 (CD-ROM)
19. Niwa Y, Kawahito Y, Kubota S, Katayama S (2008) Proc ICALEO 2008, LIA 101: 311–317 (CD-ROM)
20. Katayama S, Jung K-W, Kawahito Y (2010) Proc 29th ICALEO 2010, LIA 103:333–338 (CD-ROM)
21. Nagatsuka K, Sech Y, Nakata K J-Stage https://www.jstage.jst.go.jp/article/jspmee/3/1/3_10/_pdf

Chapter 10
Industrial Applications of Laser or Hybrid Welding

10.1 Steel Industry

In the steel industry, slabs are made of continuous casting process, then their thicknesses are thinned through hot rolling, and hot rolled steel plates are manufactured through acid pickling and annealing. The thicknesses of the hot rolled coils are furthermore thinned, and cold rolled steel sheets are produced through acid pickling, annealing, and surface treatment. Steelmaking processes and industrial application examples of laser welding are shown in Fig. 10.1 [1]. Presently, with the objective of improving productivity and stabilizing quality, the end and the start of the coil sheets are continuously welded with laser, as the laser welding machine is shown in Fig. 10.2 [2].

Laser welding is continuously operated every several minutes on average for 24 h. The cycle time is within 2 min to secure high productivity, full automation is feasible, and the properties of welded joints are improved. Such laser welding can be applied to stainless steels and high-alloy steels, and there are many merits. In welding of high-alloy steels, the weld metals and HAZ are hardened and sometimes brittle, and therefore, a filler wire is used to reduce hardness and to prevent cracking.

In the initial stages, high-power CW CO_2 laser was used, but recently high-power fiber and disk lasers have been utilized. Spattering is likely to occur in these welding, and thus, the development of laser beam modes is actively performed by two core fibers (ARM laser) and the modification of focusing optics.

Moreover, stainless steel sheets and so on are subjected to roll forming, and long pipes are manufactured by laser welding in the longitudinal direction, as shown in Fig. 10.3 [2]. Such laser welding is tried to apply other materials.

S. Katayama, *Fundamentals and Details of Laser Welding*,
Topics in Mining, Metallurgy and Materials Engineering,
https://doi.org/10.1007/978-981-15-7933-2_10

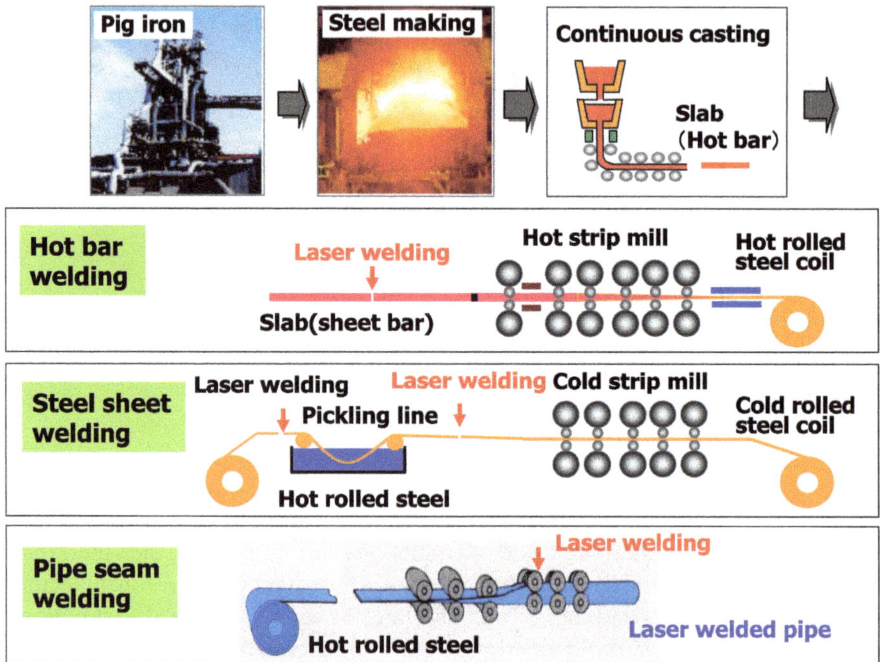

Fig. 10.1 Steelmaking processes and industrial application examples of laser welding

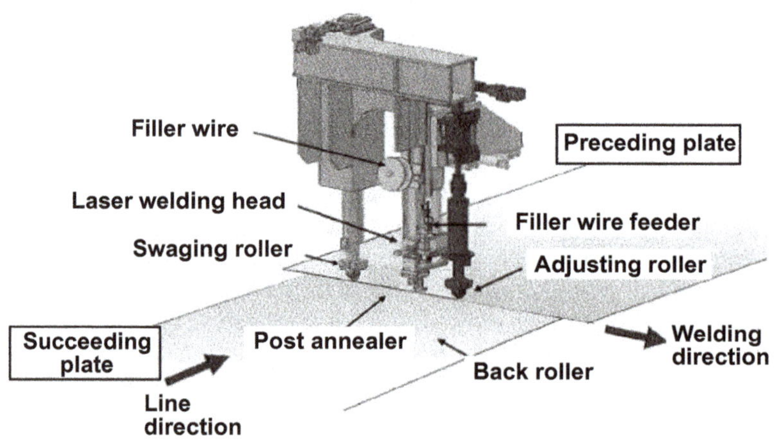

Fig. 10.2 Laser welding machine for coil joining in manufacturing line of steel sheets

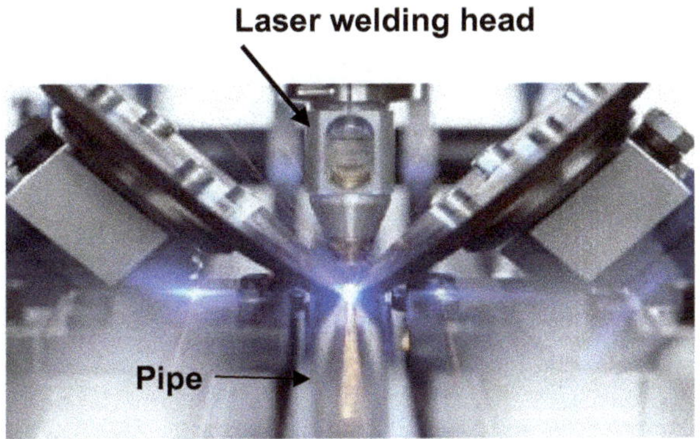

Fig. 10.3 Laser welding situation for continuous manufacturing of long pipes in longitudinal direction

10.2 Automobile Industry

In the automobile industry, laser welding including plastic joining has been applied in the initial stages of the development of respective lasers. Laser tailored blank welding started for making of large floors in the middle of 1980', and presently is employed to car bodies and doors in the world. This application is spread to other parts such as tubes and tanks. The recent lasers used are high-power and high-quality fiber, disk, and diode lasers. The kind of steels used is also increasing from various steels to 1500 MPa HT steels or Al alloys.

3D joining of Zn-coated steels or Al alloys in car bodies is performed by welding or brazing with high-efficient solid-state lasers, or laser–arc hybrid welding, as shown in Fig. 10.4 [3]. CO_2 laser joining of automotive parts started as welding of alternators and stator cores, and electromagnetic clutches for air conditioners in about 1986 in the place of arc or electron beam welding. In dissimilar joining of graphite cast iron and low alloy steel, bolt fastening or electron beam welding has been used in Japan, but in other countries, laser welding has been employed. And now the adoption of high-efficient laser welding is under consideration, as shown in Fig. 10.5 [3]. In this welding, hard and brittle ceramics (Fe_3C cementite) are formed in a cast alloy, and consequently quenching cracking is likely to occur. Therefore, pre-heating or post-heating treatment or the use of a filler wire with a high content of Ni should be needed for the guarantee of the prevention of quenching cracking.

Remote laser welding has been practically employed in conjunction with the improvement of laser beam quality and the peripheral technology. Thereby high productivity and labor saving are realized in the place of resistance spot welding.

10.3 Application to Train and Aircraft

Recent railroad vehicles and electric trains are made of chiefly austenitic stainless steel or Al alloys and partly steel, as shown in Fig. 10.6 [4].

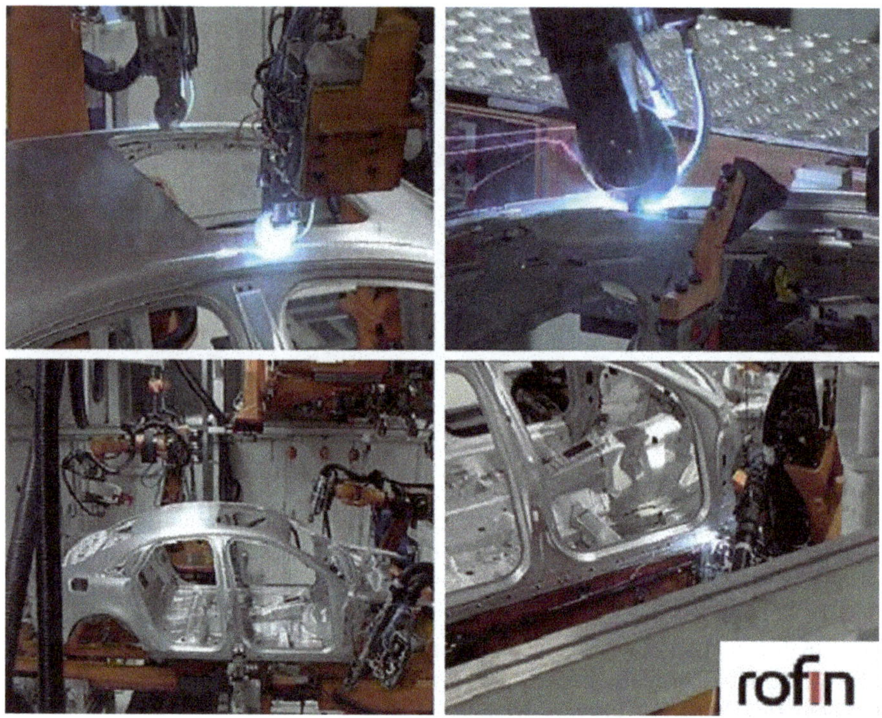

Fig. 10.4 Laser welding situations in manufacturing process of car bodies

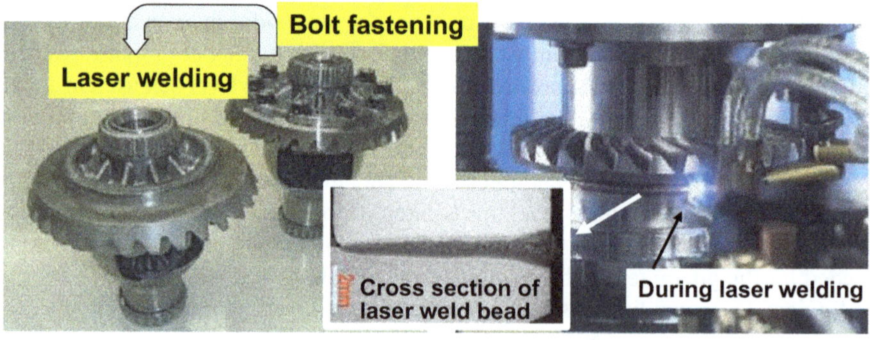

Fig. 10.5 Laser welding and its situation of cast iron and low alloy steel in place of bolt fastening

In manufacturing stainless steel vehicles, resistance spot welding has been employed, but strain-induced distortion is high, and complicated stress relieve working is required after welding, and traces or marks of spot welds are seen from the outside. On the other hand, a fiber or disk laser beam is irradiated (on the upper sheet surface) from the car inside by using a moving focusing head with rolling pressing

Fig. 10.6 Electric trains subjected to applications of laser welding

jig to produce partial penetration lap welds in stainless steel Type 304 or 301 sheets of 1–3 mm thickness. Fiber laser lap welding is performed at the power of 3 kW and the speed of 5–6 m/min. Examples of a laser partial penetration weld are exhibited in Fig. 10.7 [4]. As a result, the distortion of the joint is small, and traces or marks of laser welds are not noticeable from the outside.

In the manufacturing of Al alloy trains, fiber laser (forward), and MIG arc (following) hybrid welding was applied to produce a 25 m long weld in Al alloy A6N01-T5 extruded plate, as shown in Fig. 10.8 [5]. Good weld beads can be

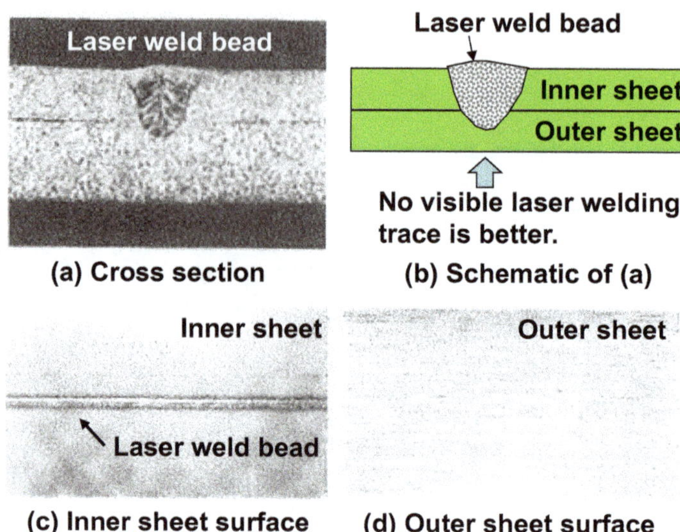

Fig. 10.7 Example of surface appearances and cross sections of laser partial penetration weld bead for electric train of austenitic stainless steel

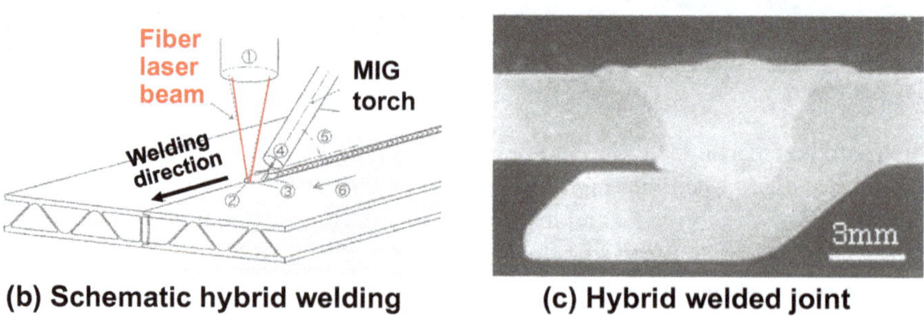

Fig. 10.8 Fiber laser (preceding) and MIG arc (following) hybrid welding system, welding situation, and hybrid welded joint for electric trains of aluminum alloys

produced. Other companies employ friction stir welding or MIG arc welding for the joining of Al alloys in trains.

Besides, in manufacturing steel trains, MAG or CO_2 gas arc welding is used.

In manufacturing aircraft, generally, adhesives for structure, rivet bolt, and pinning are chiefly used, and fusion welding has not been applied. However, recently, a slab-type high-power and high-quality CO_2 laser has been developed, and accordingly T-type fillet welding of skins (outer plates) and stringers (reinforcement plates) of Al alloy A6013 was carried out by using two sets of 3.5 kW CO_2 lasers for abdominal panels of fuselages of Airbus A380, etc. The example has been already exhibited in Fig. 7.4 [6]. Examples of airplane fuselages, skins and stringers, and laser welding situations are also shown in Fig. 10.9 [6]. Two laser beams are simultaneously irradiated from two directions. Two wires with a high content of Si are used to prevent solidification cracking since A6XXX alloys employed are susceptible to solidification cracking. Si emission in plumes is always monitored to confirm melting of wires, as described in the Sect. 7.2.

Moreover, Ti–6Al–4V Ti alloy is used for the fan cases of jet engines of airplanes, and now welding with a fiber laser is employed.

Fig. 10.9 Examples of airplanes, fuselage, skin and stringers, and laser welding situation

10.4 Application to Shipbuilding and Bridge

The thermo-mechanical control process (TMCP) is developed. And thus, rolled high tensile strength (HT) steels for welded structure are used for shipbuilding, architecture, and pipelines, and weather resistant steels are employed for bridges. Generally, submerged welding, and MIG/MAG and CO_2 gas arc welding are used for these structures, and recently the application of laser and arc hybrid welding is investigated to deal with plate gaps of existing structures.

In Europe, CO_2 laser and MAG arc hybrid welding was first applied to weld butt joints, and then fiber or disk laser and MAG arc hybrid welding are used for butt joints and fillet joints in luxury cruise ships. An example of ship, fiber laser and MAG arc hybrid welding situation, and hybrid weld made in X70 steel plate of 12 mm thickness at 10.5 kW and 2.2 m/min are shown in Fig. 10.10 [7, 8]. In Japan, hybrid welding using fiber laser and CO_2 gas arc was applied for general merchant ship [9]. The merits of applications of hybrid welding are (1) reduction of post-process, (2) cost reduction due to the decrease in construction man-hour, (3) shortening of the construction period, and (4) weight-saving due to the reduction in deposited metal amount. When the plate is thick, burn-through, and welding defects such as underfilling and undercutting is likely to occur. A backing ceramic plate may be used to prevent such welding defects.

Bridges of C–Mn steel or HT steel plates were tired to be manufactured by fiber laser and MAG hybrid welding, as shown in Fig. 10.11 [10]. Incidentally, a monitoring and tracking system was also developed for the production of long weld beads.

Fiber or disk laser and MAG arc hybrid welding were also tried to establish the welding process for pipelines, as shown in Fig. 10.12 [7, 11, 12]. It is confirmed that hybrid welding is feasible in the on-site production of pipelines by using the welding machines on the movable trailer.

Fiber delivery of YAG laser, disk laser, and fiber laser can be adequate for the application to underwater welding, as shown in Fig. 10.13 [13]. Underwater welding were also tried, and a double gas nozzle may be good for expelling water from the welding part and performing stable laser welding under water conditions.

10.5 Electrical and Electronic Industries

The electrical industry is a manufacturing field of refrigerators, lighting appliances, batteries, generators, telephones, and so on. The electronic industry is a manufacturing field of television (TV), digital cameras, mobile or portable phones, control equipment, transistors, integrated circuit (IC), computers, and so on. A variety of lasers are used for various manufacturing parts and goods. Laser drilling and cutting are the main processing, though.

(a) Luxury cruise ship

(b) Hybrid welding for floors (c) Hybrid weld

Fig. 10.10 Example of luxury cruise ship, fiber laser and MAG arc hybrid welding situation, and hybrid weld made in X70 steel plate of 12 mm thickness at 10.5 kW and 2.2 m/min

The cases of Li–ion batteries of mobile phones are made of Al alloy A3003, and in order to weld a cover or a lid and a case and to seal-weld caps after liquid injection, spot and seam welding with a pulsed YAG laser or high-speed welding with a remote continuous wave (CW) fiber laser is performed at low heat inputs.

In the early stages, CO_2 laser welding was applied to frame electric motors, while pulsed YAG lasers have been used to weld not only battery cases, but also Al rotors

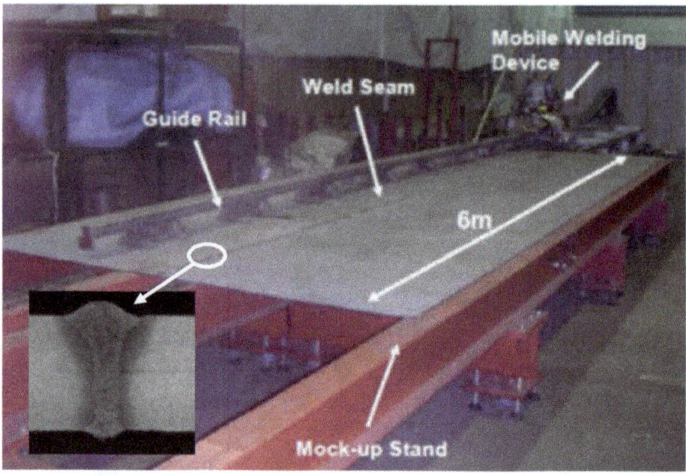

Fig. 10.11 Fiber laser and MAG arc hybrid welding system for shipbuilding and bridges

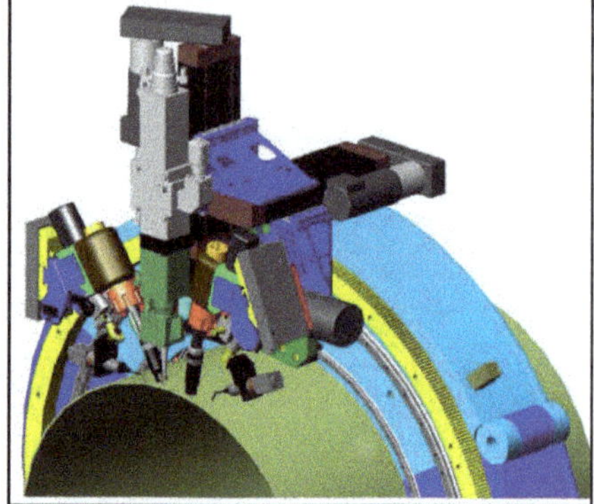

Fig. 10.12 Fiber or disk laser and MAG arc hybrid welding machine for pipelines

of small motors, electronic connectors, halogen lamps, and power sensor modules, as shown in Fig. 10.14 [14].

Recently, welding of copper (Cu) is increasing, and blue laser, green laser, adjustable ring mode (ARM) fiber laser, etc., are tried to weld Cu sheets or dissimilar Cu–Al sheets for motors and batteries.

10.6 Jewelry, Glass Frame, and Medical Industry

In the jewelry industry, spot welding with a pulsed YAG laser is used for the production of platinum, gold, or silver rings, as the spot-welding situation is shown in

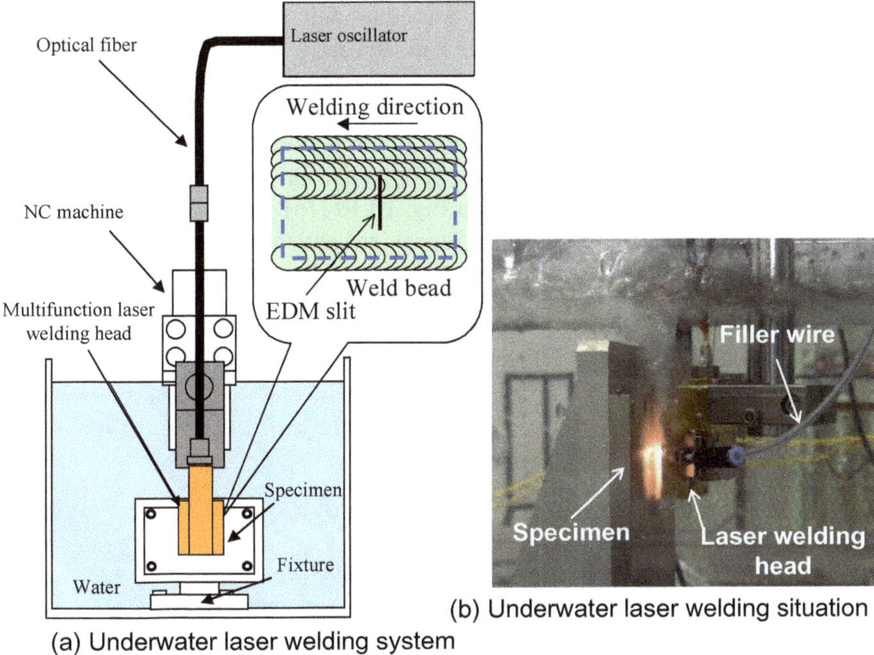

Fig. 10.13 Underwater welding system and welding situation

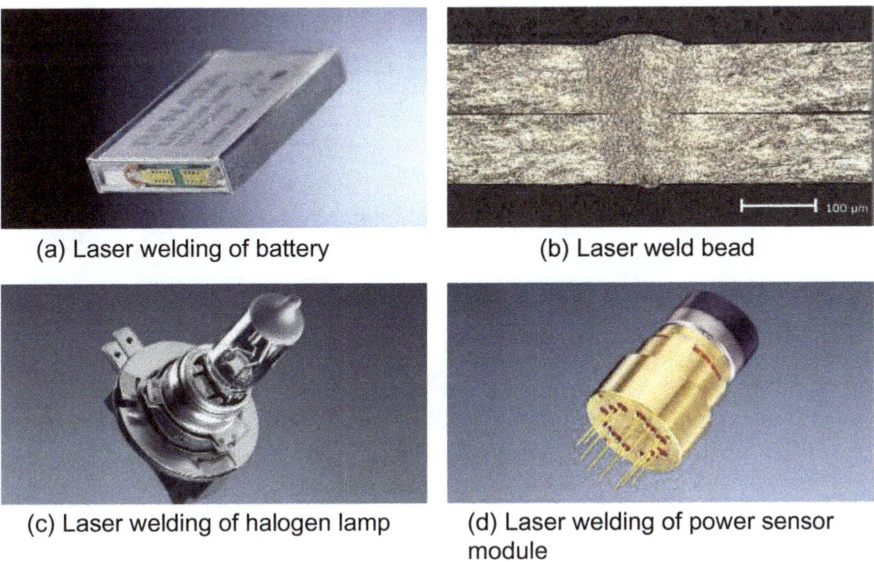

Fig. 10.14 Examples of various applications of laser welding in electrical and electronic industries

Fig. 10.15 [15]. The pulsed YAG laser is used to melt the ring nail for the fixing of a diamond by shooting its beam on the nail. Recently, the use of continuous wave (CW) or pulsed wave (PW) fiber lasers is increasing because of the production of smaller weld fusion zones.

In the glass frame industry, spot welding with a pulsed YAG laser is developed for the production of Ti alloy, and laser spot and seam welding is practically used by repeating spot welding, as shown in Fig. 10.16 [16]. In conventional resistance spot welding, weld traces or marks are recognized and the mechanical properties of the welded joints are degraded. However, laser welds are hardly degraded and the

Fig. 10.15 Laser welding for jewelry

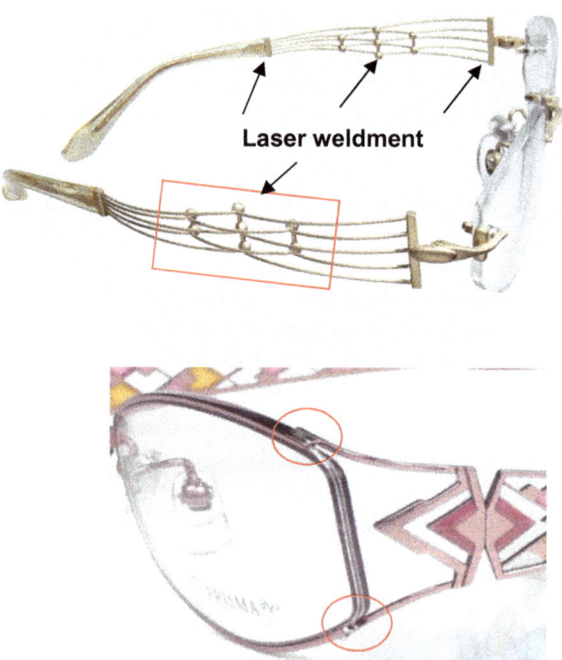

Fig. 10.16 Examples of glass frames welded with pulsed YAG laser

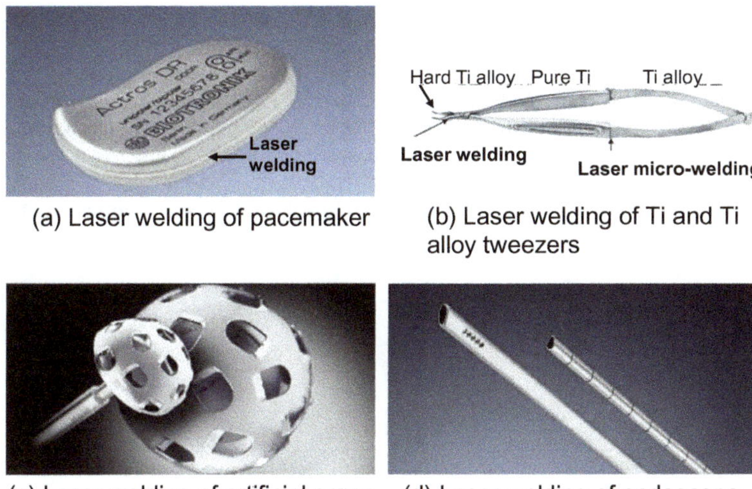

(a) Laser welding of pacemaker (b) Laser welding of Ti and Ti alloy tweezers

(c) Laser welding of artificial organ (d) Laser welding of endoscope

Fig. 10.17 Pacemakers, tweezers, artificial organ, and endoscope made by laser welding

properties of welds are improved. Moreover, seam welding with a CW fiber laser is also investigated as a high-speed production process.

In the medical industry, spot or seam welding with a PW or CW laser is applied to produce sound weld beads in pacemakers or surgical scissors, tweezers, artificial organ, endoscope, and knives, as some examples are shown in Fig. 10.17 [14, 16]. In welding of small parts, pulsed YAG or fiber lasers are still chiefly used.

References

1. Ono M (2015) JFE steel, personal communication
2. Kitani Y, Miyata J (2018) JFE steel and tadadenki, personal communications
3. Beyer E (2018) (Fraunhofer IWS), Personal communication, laser market place 2003
4. Hirashima T, Umebayashi T, Taguchi M (2006) Tech. Rep. Kawasaki Heavy Ind. 160:50–53 (in Japanese)
5. Yoneya H (2008) J. Light Metal Welding, JLWA 46:43–47 (in Japanese)
6. Schumacher J, Zerner I, Neye G, Thormann K (2002) In: Proceedings of ICALEO 2002 (laser materials processing conference), LIA, vol 94, Session A-Welding (CD)
7. Scherbakov E, Kikuchi J (2019) Personal communication, IPG photonics corporation. Web (https://www.ipgphotonics.com/)
8. Litmeyer M, Lembeck H (2008) In: Proceedings of the 70th laser materials processing conference, JLPS, pp 139–143
9. Koga H (2008) Proceedings of LMP symposium, JWES, LMP committee, pp 139–143 (in Japanese)
10. Inose K, Kanbayasi J, Ido N, Oowaki K, Miyaji T (2009) Tech. Rep. IHI 49(1):60–63
11. Gumenyuk A, Rethmeier M (2013) Developments in hybrid laser-arc welding technology. In: Katayama S (ed) Handbook of laser welding technologies, Woodhead Publishing, pp 505–521

12. Keitel S (2010) Proceedings of the 8th international conference on beam technology, SLV Halle, Germany, pp 8–15
13. Makino Y (2015) Research Result in Toshiba Corp., Personal communication
14. Nakamura T (2018) The latest laser welding system of TRUMPF. In: Proceedings of the 89th laser materials processing conference, Osaka, Japan, vol 89 pp 55–60 (in Japanese)
15. LaserStar Technologies Corp. (Website Homepage) (2018, 2020). https://www.laserstar.net/en/products/
16. Nakamura H (2011) In; Proceedings of the 75th laser materials processing conference, JLPS, pp 39–43 (in Japanese)

GPSR Compliance
The European Union's (EU) General Product Safety Regulation (GPSR) is a set of rules that requires consumer products to be safe and our obligations to ensure this.

If you have any concerns about our products, you can contact us on

ProductSafety@springernature.com

In case Publisher is established outside the EU, the EU authorized representative is:

Springer Nature Customer Service Center GmbH
Europaplatz 3
69115 Heidelberg, Germany

www.ingramcontent.com/pod-product-compliance
Ingram Content Group UK Ltd.
Pitfield, Milton Keynes, MK11 3LW, UK
UKHW021011290726
14059UKWH00001BA/70
* 9 7 8 9 8 1 1 5 7 9 3 2 5 *